Parrots: *The Animal Answer Guide*

Parrots

The Animal Answer Guide

Matt Cameron

The Johns Hopkins University Press Baltimore

BOWLING GREEN STATE
UNIVERSITY LIBRARIES

© 2012 The Johns Hopkins University Press
All rights reserved. Published 2012
Printed in the United States of America on acid-free paper
9 8 7 6 5 4 3 2 1

The Johns Hopkins University Press
2715 North Charles Street
Baltimore, Maryland 21218-4363
www.press.jhu.edu

Library of Congress Cataloging-in-Publication Data

Cameron, Matt, 1964–
Parrots : the animal answer guide / Matt Cameron.
p. cm.
Includes bibliographical references and index.
ISBN 978-1-4214-0543-8 (hdbk. : alk. paper) — ISBN 978-1-4214-0544-5 (pbk. : alk. paper) — ISBN 978-1-4214-0659-6 (electronic) — ISBN 1-4214-0543-1 (hdbk. : alk paper) — ISBN 1-4214-0544-X (pbk. : alk. paper) — ISBN 1-4214-0659-4 (electronic) 1. Parrots—Miscellanea. I. Title
QL696.P7C319 2012

598.71—dc23 2011047315

A catalog record for this book is available from the British Library.

Special discounts are available for bulk purchases of this book. For more information, please contact Special Sales at 410-516-6936 or specialsales@press.jhu.edu.

The Johns Hopkins University Press uses environmentally friendly book materials, including recycled text paper that is composed of at least 30 percent post-consumer waste, whenever possible.

To Christine, Katelyn, and Alana

I was awakened at 5 a.m. by a whirring noise, and looked out of the sleep-out to see Shellies [Budgerigars] in millions. They went on with hardly a break until about 8 a.m. They were like clouds at times, and actually darkened the sky, a very wonderful sight. An estimate of the number would be impossible, but I could say that they were passing for three hours, and any number from one to ten millions must have passed every ten minutes.

Byron McLachlan, South Australia, 1932

Contents

Acknowledgments

This book could not have been written without access to the published findings of dedicated parrot biologists. Despite the many demands on their time, they always responded positively to my requests for assistance. I am especially grateful to those who reviewed draft answers and chapters. They include Kathryn Arnold, Matt Berg, Robin Bjork, Jack Bradbury, Don Brightsmith, Jessica Eberhard, Suzanne Eckert, Terry Greene, Andrew Mack, Juan F. Masello, Ron Moorhouse, Penny Olsen, Irene Pepperberg, Camilla Ribas, Craig Symes, and Tim Wright. Others answered specific questions during the drafting stage. They include Michael L. Avery, Andy Bennett, Matt Berg, Rob Heinsohn, Victor G. Hurley, Leo Joseph, Roland Kays, Kevin McGraw, Richard Prum, and Devi Stuart-Fox.

I was fortunate in being able to call on a number of excellent photographers to enliven the text. A special debt is owed Heinz Lambert and Roy Toft (www.toftphoto.com), working photographers who provided a large number of images. Others who supplied photos include Mark Bittner, Lindsay Cupper, Cathy Katsoolis, Patrick Kelly, Tony Kirkby, Steve Milpacher, Phil Perry, Irena Schulz, and Klaus Uhlenhut. Many researchers took the time to sort through their archives and supply photos. They include Michael L. Avery, Michael Barth, Robin Bjork, Walter Boles, Suzanne Eckert, Jonathan Ekstrom, Terry Green, Victor G. Hurley, Dean Ingwersen, Arlene Levin-Rowe, Justin Marshall, Juan F. Masello, Ani Patel, Andrew Somerville, Craig Symes, Chris Tzaros, and Tim Wright.

I thank Leo Joseph for providing access to parrot skins in the Australian National Wildlife Collection and David Gowland for permitting photography of birds at the Priam Psittaculture Centre. Rosella Foods Pty. Ltd. granted permission for the reproduction of their famous parrot logo, and Australia Post granted permission to reproduce the Australian parrot series stamps and first day cover. Staff at the National Library of Australia and Australian War Memorial efficiently processed requests for access to material in their picture collections.

The writing of this book has taken some time and I am grateful for the patience shown by the team at the Johns Hopkins University Press. The support and guidance provided by Vincent Burke, my editor, was much appreciated, while Jennifer Malat ensured the publishing process ran smoothly. Sitting on my desk are well-thumbed copies of published titles in the Animal Answer Guide series. They were a valuable guide on how to approach particular topics. Over the years, discussions with a number of people have fostered my interest in parrots and given me the confidence to study and write about this amazing group of birds. They include Stephen Garnett, Terry Greene, Rob Heinsohn, Damon Oliver, Penny Olsen, and Rick Webster.

Finally, this book could not have been completed without the love and support of my wife, Chris, and daughters, Katelyn and Alana.

Introduction

My early interest in natural history was piqued not by parrots but by birds of prey. As a teenager, I recall spending hours poring over Tom Cade's 1982 *Falcons of the World* and Jack and Lindsay Cupper's 1981 *Hawks in Focus*. My dream was to become a raptor biologist. Growing up in the Riverina town of Griffith, it was parrots, however, that formed a constant backdrop to my life. Our house was only a couple of streets back from a bush-covered hill, home to a flock of Pink Cockatoos *(Lophochroa leadbeateri)*. We referred to them as "Major Mitchells" or "Wee Jugglers," though you rarely hear the latter name anymore. Small flocks regularly flew over our house, their quavering calls causing us to race outside to catch a glimpse. I'm happy to report the birds are still there 30 years later and, when I'm home visiting Dad, their calls still draw me outside. It was perhaps no surprise that I chose to study Glossy Cockatoos *(Calyptorhynchus lathami)* for my PhD. I learned that the differences between raptors and cockatoos are not as great as one might think. Their nests are difficult to find and hard to reach, individuals can be elusive but do little when found, and both have qualities that result in complex relationships with people.

My goal in writing this book was to give people a general understanding of parrot biology. I have tried to emphasize how environmental challenges and the imperative to reproduce have shaped the appearance and behavior of species. The book is arranged thematically, each chapter dealing with a

particular aspect of parrot biology or conservation. Answers stand alone, allowing people to dip in and out of the book. This has resulted in some repetition, which I hope the dedicated reader will forgive. The literature on parrots grows each day, meaning this animal answer guide is a snapshot of what is known at the time of completion. The reader should keep in mind that our knowledge in a number of fields (e.g., color and the evolutionary relationships between species) is growing rapidly. To aid readability, I have not used the standard scientific referencing system. A selected bibliography is presented at the end of the book, with a complete bibliography available at the publisher's website (www.jhu.press.edu). I have endeavored to mention the authors of key papers, and I trust others will be pleased to see the results of their work reported herein. I hope all readers enjoy the book.

Parrots: *The Animal Answer Guide*

Chapter 1

Introducing Parrots

What are parrots?

Young children can readily pick a parrot out of a storybook lineup. What is it about parrots that make them as recognizable as dogs, ducks, and dinosaurs? Their brilliant coloration is part of the explanation. Parrots are mostly green, with splashes of red, yellow, and blue. Males and females are usually similar in appearance, though males are more brightly colored than females in a quarter of species. Looking out my window, I can see examples of both these situations. A pair of Yellow form Crimson Rosellas *(Platycercus elegans)* forages in an ornamental pear, the sexes indistinguishable from each other with their blue cheeks and red foreheads. A small flock of Red-rumped Parrots *(Psephotus haematonotus)* feeds on the lawn beneath the tree, the bright color patches of the males contrasting with the dull green plumage of the females. In a few parrots, females are more brilliantly colored than males. The Glossy Cockatoo *(Calyptorhynchus lathami)* is one such species. I sometimes search for this trusting cockatoo among the rocky rises north of my home in Albury. I locate foraging flocks by the rhythmic crunch of beak-on-wood as birds process their daily quota of Drooping Sheoak *(Allocasuarina verticillata)* cones. Sitting quietly among the feeding birds, it is easy to distinguish the yellow-headed females from the brown-headed males.

The other characteristic that helps in identifying parrots is their unique body shape. A strongly curved bill, large head, and short neck give parrots a distinctive silhouette. The parrot bill is one of the great multipurpose tools of the animal kingdom. The tip of the downward-curving upper mandible fits over the upward-curving lower mandible. The base of the upper man-

"Banksian Cockatoo" by Sarah Stone (1760–1844), National Library of Australia. Banksian Cockatoo is another name for the Red-tailed Cockatoo *(Calyptorhynchus banksii)*. In fact, this watercolor is of a Glossy Cockatoo *(Calyptorhynchus lathami)*.

dible has a fleshy covering called a cere and articulates with the skull. In concert with a muscular tongue and well-developed jaw muscles, the parrot bill provides strength and dexterity. The versatility of this feeding setup is demonstrated by the catholic diet of Sulphur-crested Cockatoos *(Cacatua galerita)* in southeastern Australia. These large white cockatoos use their bill to dig up the underground parts of plants, fell stems of ripening wheat, and tear open the leathery fruit of native pines. The other defining physical characteristic of parrots is their short, powerful legs. Their feet have two toes pointing forward and two backward, an arrangement allowing them to hold and manipulate food items. Parrots are adept climbers and often hang upside down while foraging or playing.

Our fascination with parrots cannot be fully explained by their exotic appearance. Their popularity owes as much to various interrelated behavioral traits. First among these is their social nature. Wild parrots establish strong pair bonds and spend much of their time in the company of other birds. Denied these opportunities in captivity, parrots will bond with their owners and incorporate humans into a surrogate flock. Such relationships are valued by the owners of companion birds, though the large numbers of birds in refuges attest to the potential for problem behaviors to develop. Parrots also have a relatively large brain and are known for their curios-

The Kea's *(Nestor notabilis)* strongly curved upper bill is designed for discovering and exploiting food resources. Matt Cameron

ity and capacity for learning. Such intelligence reduces the likelihood of wild birds going hungry when food is in short supply. It also assists birds in navigating their way through the sometimes-complex social networks that characterize parrot society. In this regard, a parrot's vocal ability plays an important role in establishing and maintaining relationships with other birds. The cognitive and vocal skills of parrots mean captive individuals have the ability to use human words, a trait that has contributed to their popularity as pets and ensured a prominent place in mythology, literature, and popular culture.

Parrots have long lifespans. Many people have a story about a parrot that has lived forever in a cage on their grandparents' back porch. Indeed, some cockatoos have survived in captivity for over a century. My own story relates to a Blue Bonnet *(Northiella haematogaster)* my cousins obtained from Hay on the riverine plains of New South Wales. While there is debate about the age of this bird, all agree he was more than 30 years old when he died. Not a bad effort for a smallish parrot weighing less than 100 grams. It makes sense for long-lived species to produce few young, but invest significantly in their care. Humans pursue such a reproductive strategy, increasingly so in modern times. So do large parrots, which produce one or two young a year and look after them until the following breeding season. Parrots with slow life histories are vulnerable to changes in their environment. Poaching for the pet trade or the logging of nest trees has caused populations of many species to decline to the point that parrots are now one of the most threatened groups of birds in the world.

Are cockatoos also parrots?

Yes, cockatoos are a type of parrot found in Australasia. With the exception of the Cockatiel *(Nymphicus hollandicus)*, they are large stocky birds. Their most distinguishing feature is an erectile crest, which is raised when birds are agitated or during social interactions. The shape and size of their massive bills vary depending on diet. They are highly gregarious, forming noisy, conspicuous flocks. Cockatoos are mostly white, black, or grey. The structure of their feathers means they are incapable of producing the greens and blues present in the plumage of other parrots. Their plumage is permeated by a fine powder, produced by specially modified feathers known as powder downs.

How many kinds of parrot are there?

There are approximately 360 species of parrot. The number is not fixed, as taxonomic review can result in the amalgamation or splitting of species. Scientists do not always agree on the outcomes of such reviews. For two populations to be considered distinct species, they must not be able to interbreed. This is a convenient test when populations occur together, but less helpful when they have separate distributions. In the latter circumstance, scientists look for differences between populations that would prevent interbreeding if they came into contact. Attributes commonly used to separate populations include color, size, and behavior. Defining species this way is known as the biological species concept.

Some scientists believe the biological species concept does not adequately capture the diversity present within taxa. They argue that populations warrant recognition as species if they have a recent independent history and can be differentiated on the basis of some fixed character. The potential for populations to interbreed if they come into contact is not a consideration. This approach is termed the phylogenetic species concept. If fully implemented, it would result in a large number of new species. For example, many existing subspecies would gain status as species. In addition, some populations that are indistinguishable in the field may be considered separate species where DNA evidence suggests independent evolution. For the present, the biological species concept continues to be used to classify populations to species.

Individual parrot species often have a variety of common names. For example, the Pink Cockatoo *(Lophochroa leadbeateri)* is also known as Major Mitchell's Cockatoo or Leadbeater's Cockatoo. Providing each species with a unique scientific name avoids confusion. Scientific names are comprised of two Latin or Latinized Greek words, the first of which indicates the genus and the second the species. There may be sufficient variation within a species to warrant the identification of subspecies. In this situation, a third name is added to the scientific name. The scientific name may provide clues to the appearance or behavior of species, though some are chosen to honor people or a place.

Yellow-tailed Cockatoo *(Calyptorhynchus funereus)* with its lower bill covered by facial feathers (captive).
Matt Cameron

The Red-tailed Cockatoo *(Calyptorhynchus banksii)* provides a good example of how scientific names are derived and applied to taxa. *Calyptorhynchus* comes from the Greek for "covered bill" and refers to the covering of the lower bill by facial feathers, a trait present in all members of this black-cockatoo genus. The species name *banksii* honors Sir Joseph Banks (1743–1820), who sailed with James Cook (1728–1779) on his first voyage through the Pacific. Five Red-tailed Cockatoo subspecies have been recognized, with the subspecies in southwest Western Australia known as the Forest Red-tailed Cockatoo *(Calyptorhynchus banksii naso)*. The subspecies name *naso* is Latin for "large nosed."

Where do parrots live?

In the minds of many people, parrots are closely associated with tropical rainforests. The movie industry has played a part in the development of this stereotype, with parrots forming a colorful backdrop to scenes ostensibly played out in tropical locales. The reality is very different. Although parrots are a characteristic component of tropical biodiversity, they are also broadly distributed throughout southern temperate regions. A few species have more northerly distributions. The extinct Carolina Parakeet *(Conuropsis carolinensis)* once occurred as far north as Albany, New York, while the distribution of a number of *Psittacula* species includes the Himalayas and southern China. In addition, escaped aviary birds have established self-sustaining populations outside the natural range of parrots. It is now possible to encounter parrots in the heart of major cities such as Los Angeles, New York, and London.

Central and South America. Lowland rainforests in the Neotropics support large numbers of parrots. The Amazon Basin has 40 species, with up to 20 of these occurring at a single location. In southeast Brazil, remnants of lowland Atlantic Forest are home to as many as 15 species of parrot. Lowland rainforests are made up of many different tree species, which produce abundant fruit in a variety of sizes. Large fruits are common, and these are targeted by parrots. Emergent trees are an important nesting resource, as are mature palms associated with swamps and other wet areas. Parrots partition resources between themselves, size being an important factor in determining who eats what and who nests where. If pairs of closely related species occur together, one or both will favor specific habitats. Extensive areas of lowland rainforest remain in the Amazon Basin. Elsewhere in the Neotropics, only remnants of formerly widespread lowland forests are to be found. Clearing for agriculture and logging, exacerbated by road construction, are ongoing threats. Fragmentation and degradation of forests predisposes them to fire, with frequent fire permanently altering the structure and composition of plant communities.

Mountain rainforests in the Neotropics support fewer parrots than their lowland counterparts. Extensive areas of tropical rainforest are found along the eastern slopes of the Andes, where the structure and composition of forests change with elevation. Small green parakeets belonging to the genera *Psilopsiagon* and *Bolborhynchus* are closely associated with mountain habitats in the Andes. They are collectively referred to as "mountain parakeets." Genera more typical of lowland habitats, such as *Aratinga* and *Pionus*, have mountain representatives. At the southern tip of South America, temperate rainforests provide habitat for Austral Parakeets *(Enicognathus ferrugineus)* and Slender-billed Parakeets *(E. leptorhynchus)*. Little is known about the ecology of parrots occupying mountain rainforests, though some move between elevational zones as they track food resources. In many areas, the clearing of lowland rainforests has increased the significance to parrots of mountain habitats.

The Neotropics has extensive areas of savannah and dry forest. In central Brazil, sandy soils and a strongly seasonal climate have resulted in a broad expanse of grassland and woody vegetation known as the Cerrado. Areas of continuous woodland support Red-and-green Macaws *(Ara chloropterus)*, while Hyacinth Macaws *(Anodorhynchus hyacinthinus)* favor more open habitats. Both species nest on sandstone cliffs that populate the landscape. Broadscale cropping has seen extensive clearing of the Cerrado in recent decades, with deforestation rates exceeding those in the Amazon. To the west of the Cerrado lies the Pantanal, a large wetland ecosystem that extends across the Brazilian border into Bolivia and Paraguay. Subdued topography results in widespread flooding during the wet season. Expanses

View from the canopy tower at Posada Amazonas, Tambopata River, southeastern Peru. Matt Cameron

of grassland are broken by fingers of riparian forest, while higher ground is covered with scrub and woodland. The Pantanal supports a large population of Hyacinth Macaws, birds feeding on palm nuts and nesting in tree cavities. Grazing in the Pantanal limits the regeneration of their food and nest trees, while the burning of grazing lands can result in the destruction of nest sites.

A number of Neotropical parrots with restricted ranges are associated with arid environments. In northeast Brazil, the Caatinga is a land of poor soils and limited rainfall. Scrubby vegetation provides habitat for two endemic parrots, the critically endangered Lear's Macaw *(A. leari)* and the extinct-in-the-wild Spix's Macaw *(Cyanopsitta spixii)*. The Caatinga has a long history of human exploitation. Logging, burning, and grazing have all taken their toll, with the riparian forests on which Spix's Macaw relied worst affected. In Central America, several *Amazona* species are associated with dry environments. The White-fronted Amazon *(Amazona albifrons)* is common on the Pacific slope, where it appears to have benefited from the expansion of agriculture. It also occurs on the Yucatan Peninsula, which it shares with the similarly abundant Yellow-lored Amazon *(A. xantholora)*.

The Cerrado of central Brazil provides important habitat for parrots, including cliff-nesting Hyacinth Macaws *(Anodorhynchus hyacinthinus)*. Matt Cameron

The thorn scrubs of northern Venezuela support the endemic and threatened Yellow-shouldered Amazon *(A. barbadensis)*. The dry western slopes and intermontane valleys of the Andes provide habitat for a number of parrots. The endangered Red-fronted Macaw *(Ara rubrogenys)* has a restricted range in central Bolivia, occupying dry valleys in the eastern Andes.

Africa. In Africa, a band of tropical lowland rainforest straddles the equator, extending from the Indian Ocean to the Rift Valley. This large expanse of evergreen forests contains only a few species of parrot. The best known of these is the Grey Parrot *(Psittacus erithacus)*, which historically has been heavily trapped and traded due to its popularity as a cage bird. This central African rainforest is ringed by successive belts of increasingly dry and open vegetation. Meyer's Parrots *(Poicephalus meyeri)* are broadly distributed within these drier habitats, occupying wooded country from South Africa to Sudan. The closely related Rueppell's Parrot *(P. rueppellii)* occupies arid environments in Angola and Namibia. The Rose-ringed Parakeet *(Psittacula krameri)* has a broad distribution north of the equator, sandwiched between the central African rainforest and the Sahara Desert. African mountain forests are fragmented and limited in extent. Black-winged Lovebirds *(Agapornis taranta)* live in the Ethiopian Highlands, while the Cape Parrot *(Poicephalus robustus)* occupies *Podocarpus* forests along the eastern escarpment of South Africa. Islands off the coast of Africa support several endemic species. Madagascar is home to three parrots, including the Vasa Parrot *(Coracopsis vasa)*. The endangered Mauritius Parakeet *(Psittacula eques)* is restricted to the island for which it is named.

South Asia. The Rose-ringed Parakeet is widely distributed across South Asia, including Pakistan, India, and Nepal. This species has established feral populations in other areas. The Alexandrine Parakeet *(Psittacula eupatria)* and Plum-headed Parrot *(P. cyanocephala)* are also broadly distributed, the range of the former species extending into Southeast Asia. Both species are popular with bird keepers around the world. Other *Psittacula* have more restricted distributions. The Malabar Parakeet *(P. columboides)* is endemic to the Western Ghats in southwest India, while the Slaty-headed Parakeet *(P. himalayana)* occurs in the Himalayas. A number of the *Psittacula* share habitats with hanging-parrots, either the Vernal Hanging-parrot *(Loriculus vernalis)* or Sri Lanka Hanging-parrot *(L. beryllinus)*. Hanging-parrots occur in Malaysia and western Indonesia, areas that also provide habitat for Blue-rumped Parrots *(Psittinus cyanurus)*. The parrots of South Asia have been little studied. Ascertaining their natural habitat preferences is difficult given the extent to which humans have impacted on the environment, but most appear capable of utilizing a variety of vegetation types, including cultivated areas.

Southeast Asian and Pacific Islands. Collectively, these islands support almost as many parrots as the Neotropics. Their geographic spread and complex topography include a wide variety of habitats. Areas receiving high rainfall throughout the year are dominated by tropical rainforest. Seventeen species of parrot have been recorded from a few hectares of lowland rainforest in New Guinea. Some lowland parrots extend their distribution into mountainous areas, while others are replaced at higher elevations by closely related species. On the Moluccan island of Seram, the Red Lory *(Eos bornea)* favors lowland environments, while the Blue-eared Lory *(E. semilarvata)* occupies mountainous habitats. At very high elevations, tropical forest gives way to temperate rainforest and then to grasslands. A number of parrots can be found at the treeline in New Guinea, including the Papuan Lorikeet *(Charmosyna papou)*. This mountain lorikeet has two color morphs, a red-bodied form at moderate elevations and a black-bodied (melanistic) form at higher altitudes. Where the climate is drier, lowland rainforest is replaced by monsoon forest and savannah woodland. Repeated burning helps maintain grassland habitats. On many islands, lowland habitats have been extensively cleared for agriculture. Parrots have learned to exploit cultivated foods, with some species considered agricultural pests. A reduction in logging in the west of the region has been matched by an expansion of forestry operations in the east.

Australia. The Australian arid zone dominates the center of the continent, extending to the coast in a number of places. Parrots are a major component of the desert bird fauna, both in terms of species richness and abundance. Desert parrots seek out patchily distributed food resources within commuting distance of water. The establishment of artificial water points has benefited some species, while habitat degradation linked to pastoral activities has caused others to decline. The arid center is ringed by semiarid shrublands and woodlands, which grade into temperate woodlands in the south and tropical woodlands in the north. The temperate woodlands have been extensively cleared for cropping; in some regions as little as 5% of the original native vegetation remains. Open-country parrots that readily feed on crops and agricultural weeds are at an advantage, while canopy foragers reliant on native vegetation are at a disadvantage. In the tropical woodlands, a combination of grazing and inappropriate fire regimes has reduced the food supply of granivores, causing a contraction in the range of species such as the Golden-shouldered Parrot *(Psephotus chrysopterygius).*

On Cape York in the far north of the continent, lush tropical rainforest provides habitat for species more typical of Southeast Asian islands, including the Eclectus Parrot *(Eclectus roratus)* and Red-cheeked Parrot *(Geoffroyus geoffroyi).* Tall eucalypt forests in southeast Australia are home to King Parrots *(Alisterus scapularis)* and Gang-gang Cockatoos *(Callocephalon fimbriatum),* while those in the far southwest provide habitat for Baudin's Cockatoo *(Calyptorhynchus baudinii).* Unsustainable logging practices in these areas have reduced the density of large old trees and the availability of nesting hollows. In recent decades, global warming has increased the scale and frequency of forest fires. Changes in fire regime have the potential to reduce the extent of plant communities relied on by some parrots. Australia has no true mountain parrots, though a number of species can be found at high elevations.

New Zealand. The temperate rainforests of New Zealand are home to a number of unique species. In the north and at lower elevations, conifer-broadleaf forests dominate. Podocarps, an ancient family of conifers that originated in Gondwana, are a visually striking and ecologically significant part of these forests. The fruit of these conifers, comprising a seed nestled in a brightly colored fleshy aril, is highly attractive to birds. In the south and at higher elevations, beech forests dominate. Kaka *(Nestor meridionalis)* can be found in all forest types. It is patchily distributed on the North Island and widespread west of the Southern Alps. Kaka are common on

Beech forest on the slopes around Lake Rotoiti provides habitat for South Island Kaka *(Nestor meridionalis).* Matt Cameron

predator-free offshore islands. Kea *(N. notabilis)* occur on the South Island, where they are most likely to be encountered in high altitude beech forests and alpine grasslands. Kakapo *(Strigops habroptila)* are extinct on the mainland, with intensively managed populations established on offshore islands. These favor lowland podocarp forest. *Cyanoramphus* parakeets were once common on the mainland, but species such as the Red-crowned Parakeet *(Cyanoramphus novaezelandiae)* are now more common on predator-free offshore islands. A number of *Cyanoramphus* species are endemic to Pacific and subantarctic islands. The availability of podocarp and beech seed is an important influence on the extent of breeding in large New Zealand parrots. Clearing and logging has reduced the extent and quality of forest habitat. Introduced mammals prey on adults and nestlings, while birds must compete with introduced species for critical food resources. Aviary escapes and deliberate introductions have resulted in a number of Australian parrots establishing populations in New Zealand.

What is the current classification of parrots?

Scientists use a hierarchical classification system to indicate the relationships between animals. Birds are placed in a class of vertebrates called Aves. This class is divided into orders containing groups of related species. All parrots belong to the order Psittaciformes, one of the largest orders by species number. Orders are divided into families. The Psittaciformes is a relatively homogenous group of species and was for a long time considered to have only a single family. It is now accepted that the Psittaciformes comprises at least three families—the Strigopidae, Cacatuidae, and Psittacidae. Recent molecular studies may result in the recognition of additional families.

Obvious groupings within parrot families have led to the identification of subfamilies, which are sometimes further divided into tribes. Ian Rowley identified three subfamilies within the cockatoos based on plumage color and general appearance. The Palm Cockatoo *(Probosciger aterrimus)* was placed with the black-cockatoos *(Calyptorhynchus)* in the subfamily Calyptorhynchinae, but recent molecular research has shown it to be more closely related to the white/grey cockatoos in the subfamily Cacatuinae. In the Psittacidae, the treatments have been many and varied. Nigel Collar recognized two subfamilies, the Loriinae and the Psittacinae. He divided the latter subfamily into nine tribes. The largest of these was the Arini, containing all Neotropical species. Allocation to subfamilies or tribes has traditionally been based on similarities in appearance and behavior, overlain by geography. Molecular data have been used where available. These groupings have wide acceptance, though scientists often disagree about their placement within the hierarchy. For example, the pygmy-parrots have variously been afforded the status of tribe, subfamily, and family.

Extensive molecular studies have recently confirmed some historical parrot groupings, but also shed new light on relationships between some genera and species. Distinct groupings well supported by these studies included the cockatoos, lorikeets, Neotropical parrots, and large New Zealand parrots. The *Coracopsis* parrots of Madagascar were found to be more closely related to New Guinea's Pesquet's Parrot *(Psittrichas fulgidus)* than they were to other Afrotropical species. The psittaculine (typically red-billed) group of parrots requires revision. It appears the African lovebirds *(Agapornis)* and the mostly Australasian hanging-parrots *(Loriculus)* are a distinct group, as are the New Guinea tiger-parrots *(Psittacella)*. These traditional psittaculine genera are now thought to be more closely related to the platycercine parrots (broad-tailed species centered on Australia). The platycercines require revision, though a core group that includes *Platycercus*, *Psephotus*, *Lathamus*, and *Cyanoramphus* is well supported. The Budgeri-

Table 1.1. Classification of Psittaciformes recognized by Ian Rowley (cockatoos) and Nigel Collar (parrots) in the *Handbook of the Birds of the World*, Vol. 4 (1997).

Family Cacatuidae (Cockatoos)
- Subfamily Calyptorhynchinae (black cockatoos)
- Subfamily Cacatuinae (white/grey cockatoos)
- Subfamily Nymphicinae (Cockatiel)

Family Psittacidae (Parrots)
- Subfamily Loriinae (lorikeets)
- Subfamily Psittacinae (parrots)
  - Tribe Psittrichadini (Pesquet's Parrot)
  - Tribe Nestorini (Kea and Kaka)
  - Tribe Strigopini (Kakapo)
  - Tribe Micropsittini (pygmy-parrots)
  - Tribe Cyclopsittacini (fig-parrots)
  - Tribe Platycercini (platycercine parrots)
  - Tribe Psittaculini (psittaculine parrots)
  - Tribe Psittacini (Afrotropical parrots)
  - Tribe Arini (Neotropical parrots)

gar *(Melopsittacus undulatus)* is traditionally placed with the platycercines, but molecular evidence points to lorikeets being their closest relatives. The Australian grass parrots *(Neophema, Neopsephotus)* and Australian *Pezoporus*, usually located with the platycercines, have been found to be sister taxa and likely shared a common ancestor with core platycercines.

When did parrots evolve?

Parrots have a mostly southern distribution. They are found in South America, Africa, Madagascar, India, Australia, and New Zealand. These landmasses were once part of Gondwana, leading to the hypothesis that parrots have a Gondwanan origin. Australasia has long been considered to be the center of origin, given the large number of present day genera endemic to the region. The breakup of Gondwana was thought to have led to development of separate parrot lineages, the southern continents carrying parrots with them when they split from Gondwana. Parrots were not believed to have crossed oceans to colonize continents devoid of parrots or to supplement existing parrot faunas, though a lack of fossil evidence made it difficult to determine the relative roles of fragmentation versus colonization in the evolution of regional parrot faunas.

Advances in molecular techniques have allowed scientists to examine the relationships among lineages and estimate the time they have been isolated from one another. Relationships among species are determined by the similarity of their DNA sequences. Molecular dating is based on the observation that once populations become isolated, their DNA diverges at a relatively constant rate. This divergence is due to chance mutations being fixed in a population via genetic drift, as opposed to being selected because they enhance the fitness of individuals. Genetic drift is a result of the random sampling of genetic material within a population each time it reproduces. When splits between particular taxa can be linked to geological events or fossil evidence, DNA sequence divergence rates within a lineage can be calibrated to provide estimates of divergence times. A number of recent studies have used molecular techniques to improve our understanding of the evolutionary history of the Psittaciformes.

Timothy Wright and coworkers analyzed the DNA from parrots representing 69 genera from around the world. They used the data to construct a family tree for the Psittaciformes. The base of the tree represents the common ancestor of all parrots. The first branch leads to the Kea and Kakapo, making the Strigopidae the oldest parrot family and the sister group to all other parrots. The next branch leads to the cockatoos, indicating that the Cacatuidae and Psittacidae share a common ancestor. The Psittacidae then split relatively conventionally into African, Australasian, and Neotropical species. There were a few surprises that suggested Africa may have been colonized by different lineages at different times. A close relationship was found between the African lovebirds *(Agapornis)* and Asian hanging-parrots *(Loriculus)*, while the Vasa Parrot *(Coracopsis vasa)* was only distantly related to other African parrots.

Wright and coworkers linked the split of Strigopidae from other parrots with the separation of New Zealand from Gondwana around 82 million years ago in the late Cretaceous. Using this point to calibrate divergence times, they identified a major diversification of parrot lineages around 60 million years ago. This was a time when Australia began separating from Antarctica, carrying the Australasian parrot lineages northward.

The existence of fossils from ancient parrots living in Europe during the Eocene (55 to 34 million years ago) raises questions about an earlier Cretaceous origin for modern parrots in the southern hemisphere. To explore this uncertainty, Wright estimated divergence times among modern parrot lineages using a calibration date of 50 million years ago, corresponding to a theoretical divergence date between modern parrots and ancient fossil forms from Europe. Under this scenario, a major diversification of parrot lineages took place in Australia around 40 million years ago. This is consistent with dates suggested by Nicole White and colleagues for early splits in the Psittaciformes. Colonization of the other southern continents would have required over-water dispersal. The possibility of a northern hemisphere origin for parrots with dispersal into the southern hemisphere cannot be conclusively discounted.

Manuel Schweizer and colleagues analyzed the DNA from a large sample of mostly African and Australasian species. They suggest that Neotropical lineages evolved in West Antarctica and South America, separated by a seaway from ancestral African parrots in East Antarctica. The separation of South America from Gondwana around 30 million years ago coincided with the expansion of Antarctic ice-sheets, isolating Neotropical lineages in South America. With Africa already separated from Gondwana, the ancestors of Africa's *Poicephalus* and *Psittacus* colonized the continent from East Antarctica.

Consistent with earlier studies, Schweizer and colleagues demonstrated that the common ancestor to all parrots lived in Australasia. They confirmed the close relationship between lovebirds and hanging-parrots, with the common ancestor of these groups most likely residing in Australasia. They emphasized the importance of transoceanic dispersal in the diversification of parrots. According to their analyses, they conclude that colonization of Africa by the ancestor of modern lovebirds would have required an ocean crossing. A similar scenario is suggested by the relationship between *Coracopsis* and the New Guinean Pesquet's Parrot *(Psittrichas fulgidus)*. Dispersal from Australia to Africa may have been aided by volcanic islands in the southern Indian Ocean serving as stepping stones. More recently, the *Psittacula*, *Psittinus*, and *Loriculus* are thought to have colonized the Indomalayan region from Australasia. *Psittacula* then went on to colonize Africa.

What is the oldest fossil parrot?

The oldest fossil described as a parrot was recovered from late Cretaceous (97 to 65 million years ago) deposits in Wyoming, United States. It was around 15 millimeters long and thought to be the tip of a lower bill. Its morphology was similar to that of living lorikeets, providing evidence that modern parrots were present during the Cretaceous. However, the assignment of this fossil to the Psittaciformes has been questioned by some paleontologists. Parrot fossils are usually described from only a few bones, and there is often conjecture as to whether they belong to parrots, another group of birds, or animals other than birds. The next oldest fossils come from the Eocene (55 to 34 million years ago), with a number of extinct parrot genera recognized from fossil finds in Europe and North America. These ancient parrots are considered to be the successive sister taxa of modern Psittaciformes.

If the late Cretaceous fossil from Wyoming is discounted, the oldest fossils of modern parrots date from the Miocene (23 to 5 million years ago).

A number of extinct genera have been recognized from Europe, suggesting a northern radiation of parrots. Some of these species had affinities with extant parrot groups, while others cannot be reliably assigned to modern taxa. Similarly aged parrot fossils are rare outside of Europe and are usually indistinguishable from extant taxa. Walter Boles described an incomplete rostrum from the Riversleigh deposits of northwestern Queensland. The fossil was similar in appearance to the maxilla of the extant Little Corella *(Cacatua sanguinea)* and Galah *(Eolophus roseicapillus)*. A fossil humerus recovered from Nebraska in the United States was considered identical in appearance, though somewhat smaller, to that of the extinct Carolina Parakeet *(Conuropsis carolinensis)*. It was accordingly assigned to the same genus, though this decision has been questioned in recent times.

Chapter 2

Form and Function

What are the largest and smallest living parrots?

One of my most memorable experiences with parrots was sitting on the edge of a clearing in central Brazil watching squadrons of Hyacinth Macaws *(Anodorhynchus hyacinthinus)* fly past, their blue bodies framed against a backdrop of green palms. Their plumage was breathtaking, but what stuck in my mind was their enormous size. The Hyacinth Macaw is the largest parrot species. It occurs as three isolated populations in Brazil. There are approximately 500 birds in the Amazon, 1,000 birds in central Brazil, and 5,000 birds in the Pantanal tropical wetlands of Brazil. The Pantanal population extends into neighboring Paraguay and Bolivia. Hyacinth Macaws are dietary specialists, each population relying on two or three species of palm nut. These palm nuts are extremely tough and difficult to crack. Birds employ different techniques for different types of nuts, but all methods are dependent on the size and configuration of the macaw's bill. The hardness of palm nuts and the size of the Hyacinth Macaw's bill may have increased in tandem, each species striving to gain the upper hand in an evolutionary arms race fought on the South American savannah. The large bill and associated musculature are anchored to a large head, which in turn necessitates a large body. The head remains oversized, giving flying birds the appearance of winged hammers.

The Kakapo *(Strigops habroptila)* is a wildlife documentary staple. Recently, a male bird named *Sirocco* gained worldwide notoriety when filmed attempting to mate with presenter Mark Carwardine in the BBC documentary *Last Chance to See*. Having the world's heaviest parrot attach itself to the back of your head is no laughing matter, and Carwardine suffered

The Hyacinth Macaw *(Anodorhynchus hyacinthinus)* is the world's largest parrot. Matt Cameron

a number of wounds before the chief Kakapo wrangler came to his rescue. The average weight of adult males is 2 kilograms, though individuals weighing 4 kilograms have been recorded. With an average weight of 1.5 kilograms, females are significantly lighter. The sexual difference in weight of Kakapo, one of a number of types of sexual dimorphism found in birds, suggests intense competition among males for females. The Kakapo has a "lek" mating system, characterized by males gathering and displaying at arenas where they are visited by females. A few males are responsible for most successful breeding attempts, though the process of mate selection by the females is poorly understood. Males are aggressive toward each other and will fight to the death. Large size may ensure physical dominance, which assists in gaining access to the best locations within the display arena. The calls of large males may be more desirable, and heavier males may have the stamina to call for longer periods. If large males produce more offspring, male size will increase over time.

The large size of female Kakapo may simply be a consequence of their genetic relatedness to large males. Competition among females for access to critical food resources may also be a factor. In managed populations, successful reproduction in females is related to the presence of mature Rimu

Table 2.1. Largest living parrots by length based on descriptive notes in *Handbook of the Birds of the World*, Vol. 4 (1997). Where more than one member of a genus is represented, only the largest species is shown.

Common name	Scientific name	Length (cm)	Weight (g)
Hyacinth Macaw	*Anodorhynchus hyacinthinus*	100	1435–1695
Red-and-green Macaw	*Ara chloropterus*	90–95	1050–1708
Red-tailed Cockatoo	*Calyptorhynchus banksii*	50–65	570–870
Kakapo	*Strigops habroptila*	64	950–3000
Palm Cockatoo	*Probosciger aterrimus*	55–60	550–1000
Spix's Macaw	*Cyanopsitta spixii*	55–57	296–400
Alexandrine Parakeet	*Psittacula eupatria*	50–62	198–258
Sulphur-crested Cockatoo	*Cacatua galerita*	45–55	815–975
Red-bellied Macaw	*Orthopsittaca manilata*	50–51	292–390
Vasa Parrot	*Coracopsis vasa*	50	—
Kea	*Nestor notabilis*	48	922
Masked Shining-parrot	*Prosopeia personata*	47	—
Pesquet's Parrot	*Psittrichas fulgidus*	46	690–800
Imperial Amazon	*Amazona imperialis*	45	—
Princess Parrot	*Polytelis alexandrae*	45	—
Australian King-parrot	*Alisterus scapularis*	42–43	209–275
Maroon-fronted Parrot	*Rhynchopsitta terrisi*	40–45	392–468
Eclectus Parrot	*Eclectus roratus*	35–42	355–615
Black-lored Parrot	*Tanygnathus gramineus*	40	—

(Dacrydium cupressinum) trees within their home range (the area used on a regular basis to meet daily needs). While Kakapo have overlapping home ranges, females appear to have exclusive access to core areas within their home range. Kakapo were once widespread throughout New Zealand, occupying a variety of habitats of varying quality. The ability of females to survive, but rarely breed, in suboptimal habitat would have ensured ongoing competition for high quality sites. Larger females may have been more successful in securing home ranges containing food resources necessary for reproduction. Kakapo are critically endangered and humans need to intensively manage reproductive events, making determining the part competition played in the evolution of large body size difficult.

The Buff-faced Pygmy-parrot *(Micropsitta pusio)* is the smallest parrot, 12 times shorter than the average Hyacinth Macaw and 400 times lighter than the heaviest Kakapo. It is one of six species of pygmy-parrot found in New Guinea and northern Melanesia. The habits of these incomprehensibly tiny parrots are poorly known. They have been observed feeding on lichen and fungus, their large feet enabling them to descend head-first down the trunks of trees and to hang beneath branches. Stomach contents

Parrot biologist Craig T. Symes with a Kakapo *(Strigops habroptila)*, the world's heaviest parrot. Craig T. Symes

indicate that seeds, flowers, and insects are also consumed. They excavate their nests in arboreal termitaria (termite nests in trees) or dead stumps, and these cavities are also employed as roost sites. Birds do not begin foraging until well after sunrise, with overcast conditions causing them to further delay emergence. This suggests energy inputs and outputs are finely balanced in these pocket-sized parrots.

What does a parrot's bill tell us about its diet?

There is considerable variation in the size and shape of parrot bills, with different bill forms reflecting different foraging niches. Bill characteristics that improve foraging efficiency will be strongly favored by natural selection, causing divergence in bill morphology between populations foraging on different foods. Parrots that act as pollinators typically have relatively short slender bills that allow access to nectar and pollen without damaging the flower. Species that feed on the underground parts of plants, or otherwise grub for food, have elongated upper mandibles that can be used like a pick when digging. The heavy black-cockatoo bill is well adapted for chopping into eucalypt seed capsules and the woody fruits of proteaceous evergreen shrubs, while the macaw's chisel-like lower mandible enables the splitting of large nuts.

BOWLING GREEN STATE
UNIVERSITY LIBRARIES

Red-breasted Pygmy-parrots *(Micropsitta bruijnii)* are among the world's smallest parrots. Heinz Lambert

The bills of a few species are adapted for feeding on particular foods. The Red-capped Parrot *(Purpureicephalus spurius)* has an elongated upper mandible, which appears to be an adaptation for extracting seeds from mature Marri *(Eucalyptus calophylla)* fruit. Marri seed capsules are approximately 45 millimeters long and 32 millimeters wide, with a valve securing the contents at one end. Birds are able to hack their way into green capsules, but this method of entry is ineffective against older, drier fruits. For these, birds penetrate the valve with their upper mandible and rake out the seed. The sympatric Baudin's Cockatoo *(Calyptorhynchus baudinii)* also relies heavily on Marri seed and has evolved a similarly shaped bill for similar reasons.

Do all parrots have the same type of bill?

Dominique Homberger identified two basic bill types within the Psittaciformes, which she labeled psittacid-type and calyptorhynchid-type bills. In the psittacid-type bill, the downward curving upper bill has a step in the lower surface against which the leading edge of the upward curving lower bill rests. The tongue is used to hold seeds against the step in the upper bill, with striations on the internal surface of the upper bill helping secure seeds in position. The leading edge of the lower bill is then employed like a chisel to split the seed coat and separate it from the kernel. Rotation of the seed by the tongue facilitates the complete removal of the seed coat. The configuration of the jaw, skull, and associated musculature has evolved to maximize the bite force of the mandible. Lateral movements of the lower

Table 2.2. Smallest living parrots by length based on descriptive notes in *Handbook of the Birds of the World*, Vol. 4 (1997). Where more than one member of a genus is represented, only the smallest species is shown.

Common name	Scientific name	Length (cm)	Weight (g)
Buff-faced Pygmy-parrot	*Micropsitta pusio*	8	10–15
Green-fronted Hanging-parrot	*Loriculus tener*	10	12
Green-rumped Parrotlet	*Forpus passerinus*	12–13	20–28
Orange-breasted Fig-parrot	*Cyclopsitta gulielmitertii*	13	27–34
Black-collared Lovebird	*Agapornis swindernianus*	13	39–41
Pygmy Lorikeet	*Charmosyna wilhelminae*	13	—
Madarasz's Tiger-parrot	*Psittacella madaraszi*	14	34–44
Amazonian Parrotlet	*Nannopsittaca dachilleae*	14	38–46
Lilac-tailed Parrotlet	*Touit batavicus*	14	52–72
Little Lorikeet	*Glossopsitta pusilla*	15	34–53
Guaiabero	*Bolbopsittacus lunulatus*	15	62–77
Golden-winged Parakeet	*Brotogeris chrysoptera*	16	47–80
Andean Parakeet	*Bolborhynchus orbygnesius*	16–17	48–50
Mountain Parakeet	*Psilopsiagon aurifrons*	16–19	45
Orange-billed Lorikeet	*Neopsittacus pullicauda*	18	25–40
Budgerigar	*Melopsittacus undulatus*	18	26–29
Blue Lorikeet	*Vini peruviana*	18	31–34
Edwards's Fig-parrot	*Psittaculirostris edwardsii*	18	105
Blue-rumped Parrot	*Psittinus cyanurus*	18	—
Varied Lorikeet	*Psitteuteles versicolor*	18–19	51–62
Scarlet-chested Parrot	*Neophema splendida*	19	36–44
Bourke's Parrot	*Neopsephotus bourkii*	19	39
Mindanao Lorikeet	*Trichoglossus johnstoniae*	20	48–62
Collared Lory	*Phigys solitarius*	20	71–92

bill relative to the upper bill are possible, but play only a minor role in food processing. The psittacid-type feeding apparatus is an adaptation for feeding on nuts that can be split or cracked by the application of force to preexisting weak points. It is the most common feeding apparatus among parrots.

Dominique Homberger notes that the calyptorhynchid-type bill is present in most Red-tailed Cockatoo *(Calyptorhynchus banksii)* subspecies, the Glossy Cockatoo *(C. lathami)*, and the Gang-gang Cockatoo *(Callocephalon fimbriatum)*. In calyptorhynchid-type bills, the leading edge of the lower bill is concave with prominent sharp corners. Muscles allowing lateral movements of the bill are well developed and anchored to the skull via the bony suborbital arch, the latter structure much reduced or lacking in species with a psittacid-type bill. The tip of the upper bill and the corners of the lower bill can be opposed, allowing birds to use the bill like a pair

Scarlet Macaws *(Ara macao)* have a heavy bill that allows them to open large nuts. Matt Cameron

of pliers. Food items are held in the foot, sometimes against the concave edge of the lower bill, while pods are opened or waste material surrounding seeds is removed. Seeds are then processed within the bill, the tongue playing an important role in securing and manipulating the seed. Glossy Cockatoos use the corners of the lower bill to open sheoak seed, while Gang-gang Cockatoos use the tip of the upper mandible to remove the coat from apple seeds.

How fast can parrots fly?

Most information on parrot flight speeds is derived from the fortuitous pacing of birds with motor vehicles, a methodology unlikely to provide reliable estimates. Using this approach, I have observed Glossy Cockatoos *(Calyptorhynchus lathami)* cruising at speeds of 25 kilometers per hour. Others have reported Glossy Cockatoos exceeding 45 kilometers per hour. Ian Rowley recorded Galahs keeping pace with a vehicle traveling at 70 kilometers per hour. Speed depends on bird aerodynamics and wing morphology, which have evolved to maximize the fitness of individuals. The size and shape of wings affect the way they perform in different situations. Some wing types are adapted for high speed, others for maneuverability. Wing types can often provide clues to a bird's lifestyle.

Parrot wing morphology falls between two extremes. At one end of the spectrum are species with broad rounded wings. This wing type provides good maneuverability, but delivers relatively slow flight speeds. It is common among sedentary species or those inhabiting forested environments. Typical examples include the Neotropical Amazons and many Australasian cockatoos. The sedentary Pink Cockatoo *(Lophochroa leadbeateri)* is a weak flier. It can be found in open environments, where the antipredator benefits of flocking may compensate for any increased susceptibility to predation due to low flight speeds. Parrots with long pointed wings lie at the other end of the spectrum. This wing type delivers high speeds and is expected in species that cover long distances or occupy open habitats. Such wings are possessed by the Swift Parrot *(Lathamus discolor)*, a species that breeds in Tasmania and winters on mainland Australia. Its rapid flight makes it vulnerable to collisions with glass windows or wire fences.

Many desert parrots wander over large areas to access food and water. Under these conditions, you would expect the evolution of wings that allow rapid and efficient flight. The Budgerigar *(Melopsittacus undulatus)* has relatively long pointed wings. This inhabitant of the Australian outback weighs around 30 grams and has a wingspan of 25 centimeters. Minimum and maximum flight speeds are unknown, though wind tunnel testing of pet Budgerigars has occurred between 19 and 48 kilometers per hour. These tests showed the most efficient (maximum distance per unit of energy) flight speed for Budgerigars was 42 kilometers per hour. The long wedge-shaped wings of the Cockatiel *(Nymphicus hollandicus)* are similarly well adapted for undertaking long-distance movements. The Galah *(Eolophus roseicapillus)* has a narrow pointed wing, especially when compared to relatives having more sedentary lifestyles. The Galah's enhanced flying ability appears not to have come at the expense of maneuverability, it being a well-known acrobat and aerial clown.

Are there any flightless parrots?

The Kakapo *(Strigops habroptila)* is the only flightless parrot. It is one of a number of flightless New Zealand birds, including several species of extant kiwis *(Apteryx)* and an extinct group of small wrens (e.g., Stephens Island Wren *Traversia lyalli)*. Flightlessness is common among island birds and evolved in the absence of mammalian predators. Growing and maintaining the structures necessary for flight requires considerable energy. When flight does not confer any survival advantage, natural selection will favor a reduction in flying ability.

Kakapo morphology reflects its flightless lifestyle. The keel of the sternum, to which the flight muscles are anchored, is reduced in size. The pectoral muscle mass to body weight ratio is the lowest of any parrot. Powerful legs enable Kakapo to walk long distances, with some individuals covering several kilometers a night. Using their legs and bill, birds climb vertical tree trunks and move through the canopy from one tree to another. Kakapo have the smallest wings of any parrot relative to their body size, which they use for balance when climbing trees or traversing rocky terrain. Birds descend from the canopy using wing-assisted leaping or parachuting, dropping a few meters at a time from higher to lower branches. Flightlessness imposes few constraints on Kakapo in terms of fat storage, allowing them to store energy when food is abundant. Fat reserves contribute to successful reproduction and assist with temperature regulation (thermoregulation) in cold environments.

A number of parrots have a mostly terrestrial existence. The Ground Parrot *(Pezoporus wallicus)* is a midsized green parrot. It has a patchy distribution on mainland Australia, but is widely distributed throughout southwest Tasmania. It lives in low dense vegetation, typically coastal heathlands, but also button-grass plains at higher elevations in Tasmania. Ground Parrots are active during the day, foraging on the ground and in low shrubs. They avoid detection by walking away through the undergrowth, but if surprised will take to the air briefly before dropping to the ground. Ground Parrots engage in bouts of flying and calling at dawn and dusk. Such displays may have a territorial function or serve to maintain pair bonds. Nests are located on the ground, sheltered by overhanging vegetation. The Night Parrot *(P. occidentalis)* is a dumpy green parrot that is reminiscent of a miniature Kakapo and shares its nocturnal habits. Like its close relative the Ground Parrot, it forages and nests on the ground.

Do a parrot's feathers wear out?

Wear-and-tear on a parrot's feathers means they must be periodically replaced. Wing feathers are commonly moulted in a systematic manner, the moult commencing in the middle of the wing and progressing inward and outward from this point. This pattern means growing feathers are protected from damage by the presence of older feathers on their outside. The moult of tail feathers is more erratic. Smaller parrots may have as many as four wing feathers growing at once, while larger parrots may only grow a single wing feather at a time. Smaller parrots will renew all their feathers at least once each year and typically complete the moult within six months. The complete replacement of feathers in large parrots can take up to two years, with birds likely to be moulting some feathers throughout the year. Body feathers are moulted coincidentally with flight feathers. The Budgerigar *(Melopsittacus undulatus)* and some other parrots are unusual in that a new cycle of moult may commence before the previous cycle is completed. In Budgerigars, such stepwise moulting is a possible adaptation to life in the deserts of Australia. The slow and continuous replacement of feathers spreads the energetic costs of moulting throughout the year and allows birds to breed whenever conditions are optimal.

A parrot's feathers are critical for flight and the regulation of body temperature. They may also be used for concealment or advertisement. Accordingly, birds spend a lot of time on their maintenance. Nibbling cleans the feathers and realigns the feather barbs. It also assists with the distribution of preen oil sourced from the preen gland on the lower back. Preen oil may be supplemented by a fine powder produced by specially modified feathers known as powder downs. Powder downs are well developed in cockatoos, but may be sparse or absent in other parrots. The preen gland is lacking in a number of Neotropical genera, with these parrots typically possessing powder downs. It is rare to see parrots bathing in pools of water, though some species deliberately wet their plumage when it rains. Despite the attention they receive, feathers wear out and must be replaced in order to maintain flight performance and thermoregulatory capacity. Feather degradation can also result in subtle changes in feather color, which may negatively impact on an individual's ability to compete for limited resources. In many parrots, replacement of feathers is necessary for birds to signal their age or attainment of sexual maturity.

The replacement of feathers requires a great deal of energy. Parrots must channel resources into the growth of new feathers at a time when feather loss makes flight and thermoregulation more energetically demanding. The moult of wing feathers impairs flying ability, which reduces foraging efficiency and increases predation risk. To reduce energy demand, moulting is separated from activities such as reproduction. In temperate Australia, parrots commence moulting toward the end of the breeding season. This allows new feathers to be grown over summer, a time when food remains relatively abundant. On the Northern Tablelands of New South Wales, adult Eastern Rosellas *(Platycercus eximius)* begin moulting primary wing feathers in early January and finish in late May. If the moult has not been completed by this time, birds defer the moult of the outermost wing feathers to the following season. The migratory Swift Parrot *(Lathamus discolor)* renews its plumage between the end of the breeding season and before leaving Tasmania for the Australian mainland. This ensures birds have fresh plumage for the crossing of Bass Strait and are well equipped for their sometimes extensive wanderings across the southeast of the continent in search of food.

Are there any bald parrots?

There are a number of bald-headed parrots. In the same way the featherless vulture head is an adaptation for feeding on bloody carcasses, bald-headedness in parrots may be an adaptation for feeding on sticky fruit. The most distinctive of the bald-headed parrots is Pesquet's Parrot *(Psittrichas fulgidus)*, a large black-and-red parrot found in the foothills and lower mountains of New Guinea. Pesquet's Parrots are not completely bald, the featherless black skin of the face ending behind the eye. Paul Igag studied Pesquet's Parrots with the assistance of the indigenous Pawaia people. They found its diet was dominated by the pulp of several fig species, providing circumstantial evidence for the benefits of bald-headedness in parrots that specialize on "sticky" foods. If avoiding soiled and matted head feathers is a significant advantage to individuals, it is surprising bald-headedness is not more widespread among other fruit-eating parrots.

There are two species of bald-headed parrot endemic to the Amazon Basin in Brazil. The Vulturine Parrot *(Pyrilia vulturine)* has a black-skinned head with a collar of yellow and black feathers, while the Bald Parrot *(P. aurantiocephala)* has an orange-skinned head and no feathered collar. Juveniles have green feathered heads, with adult skin color acquired when head feathers are molted. The Bald Parrot, sometimes referred to as the Orange-headed Parrot, was initially thought to be an immature condition of the Vulturine Parrot. It was described as a distinct species in 2002 by Renato Gaban-Lima and colleagues. The Bald Parrot is restricted to a relatively small area between the Madeira and Tapajós Rivers, its range encompassed by that of the more widely distributed Vulturine Parrot. Bare-headedness in the Vulturine Parrot has been interpreted as an adaptation for feeding on large juicy fruits. However, skin color may serve as a conspicuous and effective signal of a bird's quality. The color of bare parts can change more rapidly than feather color, potentially providing a more reliable signal of current condition. Bare part coloration may also provide information not communicated by feather color.

Another parrot displays temporary bald-headedness. On Madagascar, the female Vasa Parrot *(Coracopsis vasa)* looses her head feathers during the breeding season. This may prevent them becoming soiled when males re-

Yellow-collared Macaws *(Primolius auricollis)* have an area of bare skin on the face, a characteristic shared with other macaws (captive). Matt Cameron

gurgitate food to females, but this does not explain why the orange color of the skin becomes more intense. Vasa Parrots have a polygynandrous mating system, unusual among parrots. Females mate with several males, a number of whom may share paternity of nestlings. Males also mate with a number of females, supplying food for them and their young during the breeding season. Jonathan Ekstrom and coauthors propose that conspicuous head color in female Vasa Parrots evolved in response to competition among females for the food provided by males.

The macaws and many cockatoos have areas of bare skin on the head, though they cannot be described as bare-headed. At least some of these species can change the color of these bare areas by flushing them with blood. Typically, this occurs during courtship or aggressive interactions. The Blue-and-yellow Macaw *(Ara ararauna)* has a large area of white facial skin that encompasses the eye and can be flushed red. Palm Cockatoos *(Probosciger aterrimus)* have naked scarlet cheek patches, the color of which becomes more intense when birds display. Juan José Negro and coauthors reviewed facial flushing in birds. Areas of bare skin that can be flushed are highly vascularized, allowing birds to fill the tissue with blood to produce red colors. Facial flushing provides an instantaneous signal of a bird's condition, perhaps measured by the duration of flushing or the intensity of color achieved. Flushing is likely to be energetically costly and potentially damaging to the body, meaning it's an honest signal of condition. Highly vascularized skin patches that evolved to assist with heat loss are thought to have been co-opted for use in color displays.

Chapter 3

Parrot Colors

Why are so many parrots green?

Green plumage appears to have the obvious advantage of making birds less conspicuous among the forest canopy and thus less vulnerable to predators. While intuitively correct, the camouflage value of green coloration is difficult to prove. Circumstantial evidence comes from a study of Eclectus Parrots *(Eclectus roratus)* in the rainforests of northern Australia. Eclectus Parrots exhibit reversed sexual dichromatism, the red-and-blue female being more brightly colored than the green male. Intrigued by this anomaly, Rob Heinsohn and colleagues set out to find an explanation. Setting up camp in the Cape York rainforest, they spent several years unraveling the lifestyles of males and females. Males spent more time foraging than females and did not have access to the nest hollow for refuge. The green coloration of males appeared to have evolved in response to their heightened risk of predation, their plumage being less conspicuous against a leafy background compared to that of the female. Females competed with other females for nest hollows, a scarce resource. They displayed high in the canopy, where the contrast between their plumage and the green leaves made them more conspicuous.

While green plumage may serve as camouflage, demonstrating this improves survival rates is difficult. Quite a few parrots have little or no green in their plumage, and many of these share habitats with green parrots. Comparing mortality rates of green versus colorful species may shed light on whether or not green plumage reduces predation risk. However, collection of mortality data for wild parrots is difficult and requires intensive fieldwork over many years. Even if such data were available, comparisons

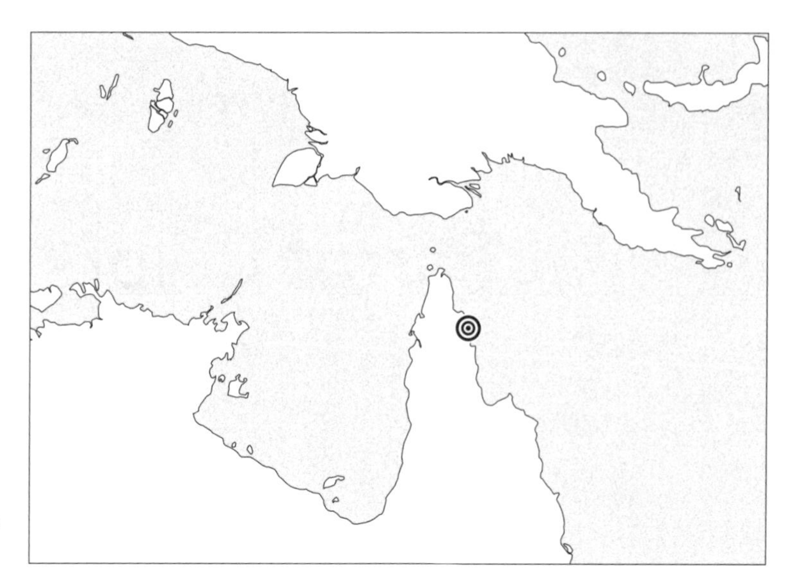

Rob Heinsohn and colleagues studied the unusual breeding system of Eclectus Parrots *(Eclectus roratus)* at Iron Range National Park on Cape York Peninsula, Queensland.

between species are confused by differences in lifestyle that may contribute to predation risk and swamp any effect of color. The same may be true for sexually dichromatic species, though studying their young provides a possible solution. Juvenile Eclectus Parrots exhibit sexual dichromatism and are thought to have similar lifestyles. A male-biased sex ratio among adult Eclectus Parrots has been attributed to higher mortality of juvenile females, who share the colorful plumage of adult females but lack access to the protection afforded by a nest hollow. This suggests that juvenile male Eclectus Parrots benefit from the camouflage provided by their green plumage.

What causes the colors of parrots?

I often flush Eastern Rosellas *(Platycercus eximius)* from the side of the road on my morning walk. This parrot is common around my house, and I rarely pay them attention. The exception is when their rainbow-like plumage is lit up by the morning sun. On these occasions, I think Eastern Rosellas must be the most dazzling of the world's birds. Early European settlers were so taken with the Eastern Rosella's appearance they referred to it as the Nonpareil Parrot or "parrot without equal." Writing in the first half of the last century, Neville William Cayley considered the Eastern Rosella to be "one of the most beautiful parrots in existence" and lamented that its abundance prevented its worth being truly appreciated. The rosella's red head and yellow abdomen are the result of colorful pigments. The blues in their wing feathers are structural colors, derived from the interaction of light with tiny structures inside the feather barbs. A combination of

Female Eclectus Parrot *(Eclectus roratus)* captured as part of Rob Heinsohn's Cape York study. Matt Cameron

pigment and structural coloration produces the pale green feathers of the rump, while melanin pigments are responsible for the black mottling on the back and wings.

The reds, yellows, and oranges in the plumage of many parrots are caused by psittacofulvins—a class of colorful pigments found only in parrots. Unlike the carotenoid pigments responsible for these colors in other birds, psittacofulvins are not consumed as part of a parrot's diet. Rather, they appear to be produced locally at maturing feather follicles. This means it is not possible to manipulate color or color intensity in captive parrots by stacking their diet with pigment-rich foods. The melanin pigments responsible for the black, brown, and grey in the plumage of parrots are shared with most other animals. The same melanin pigments are responsible for hair and skin color in humans. There are two groups of melanin pigments—eumelanins and phaeomelanins. The first group is responsible for dark black and brown hues, while the latter group is responsible for reddish-brown hues. Both groups are usually present in melanin-containing feathers. Like psittacofulvins, melanins are synthesized by the bird and are not consumed as part of the diet.

The color produced by pigments is a result of the extent to which they absorb particular wavelengths of light and the density at which they occur. Red psittacofulvins absorb most wavelengths, reflecting strongly only in the red part of the spectrum. The greater the density of red pigments, the redder the feathers appear. The subtle pinks in the plumage of some cockatoos are caused by low concentrations of red psittacofulvins. Using any objective measure, the Pink Cockatoo *(Lophochroa leadbeateri)* has to be one of the world's most beautiful parrots. Cayley considered that its delicately colored plumage made it "perhaps the most beautiful of all known species." Melanin pigments are poor reflectors. They absorb most wavelengths and

The Eastern Rosella *(Platycercus eximius)* has featured in the trademark of Rosella Foods for more than a century. Rosella Foods Pty. Ltd.

thus appear black. The relative proportion of eumelanins and phaeomelanins appears more important than the overall concentration of melanin in determining color.

The blues in parrot feathers are structural colors. These are produced by a spongy layer within the main shaft of the feather barb. This spongy layer is comprised of keratin bars and air-filled channels. Light is scattered at the keratin-air interface, with the spongy layer organized in such a way that some scattered wavelengths are canceled out while others are reinforced and reflected. This process is termed coherent scattering. Cockatoo feather barbs lack a spongy layer and are incapable of producing blue structural color, though some cockatoos have an area of blue skin around the eye (e.g., Long-billed Corella *Cacatua tenuirostris).* Blue skin is produced in a similar manner to blue feathers, with a matrix of collagen serving the same purpose as the spongy layer in feathers. Many cockatoos have predominately white plumage, a situation rare in land birds. Other parrots have white plumages patches. White is a form of structural coloration caused by the random scattering of all wavelengths of light.

Pigment and structure may interact to produce a particular color. The green in many parrots is thought to be a result of a blue structural color and a yellow pigment, with the pigment distributed throughout the cortex surrounding the spongy layer. Budgerigar *(Melopsittacus undulatus)* mutations provide a simple example of how this works. Wild budgerigars are predominately green, with black and yellow barring on the upperparts and a yellow face. In captivity, blue Budgerigars are common. These birds are the result of a genetic mutation that switches off the production of yellow psittacofulvin. As a result, feathers that were green are now blue and those that were yellow are now white. Green plumage in other bird groups is produced in a similar way, yellow carotenoids combining with blue structural coloration. Green pigments do exist, but these are restricted to a small number of birds such as touracos.

Variation in the density of melanins at various locations within the feather barb has the potential to modify structure-based and psittacofulvin-based colors. Melanin pigments often lie beneath the spongy layer. Here they absorb incident light and prevent backscattering, which has the potential to reduce the saturation of the structural color. In budgerigars, the loss of these background pigments results in pure yellow plumage. This is because light passing through the spongy layer is no longer absorbed by melanin, but reflected back through the feather. This reflected light washes out the blue structural colors to the point that only the yellow psittacofulvin is expressed. Some parrot feathers are black or brown, despite having a spongy layer. In these cases it appears that light is absorbed by melanin in the cortex before it reaches the spongy layer, preventing the expression of structural color.

The investigation of the pigment and structural coloration of parrots is in its infancy. Work by Kevin McGraw and Mary Nogare has shed light on the biochemical nature of red psittacofulvins and demonstrated their presence in a wide range of parrots having red feathers. The nature of yellow psittacofulvins remains elusive. Similarly, while melanins are undoubtedly a key determinant of parrot color, characterization of the melanin content of parrot feathers has not been undertaken. Electron microscopy has been used to examine the structure of feather barbs in only a few species of parrot. Further chemical and microscopic investigation is required before reaching definitive conclusions as to the cause of observed parrot colors.

Do parrots see the same colors as humans?

Compared to parrots, humans have limited color vision. Colors perceived by vertebrates are a function of how wavelengths of reflected light are captured by the eye and processed by the brain. Light falling on the retina at the back of the eyeball is detected by photoreceptors. The photoreceptors responsible for color vision are called cones. Humans have three types of cones, each named for that portion of the human-visible spectrum to which they are sensitive. Red cones are sensitive to long wavelengths, green cones are sensitive to medium wavelengths, and blue cones are sensitive to short wavelengths. The extent to which each type of cone is stimulated relative to the others allows the light falling on the retina to be characterized. Simply put, the eye takes three broad samples of the light reflected from an object, and the brain processes this information to allow a large variety of colors to be perceived.

Birds have four types of cones. Three of these are sensitive to wavelengths that fall within the human-visible spectrum, though each is sensitive to a narrower range of wavelengths than human cones. Oil droplets in the cones that filter certain wavelengths allow for this spectral fine tuning, the effect of which is to increase the number of colors birds can distinguish. The fourth type of cone is sensitive to ultraviolet light, which lies outside the human visible spectrum. It occurs as two types, violet-sensitive and ultraviolet-sensitive. Typically, passerines (perching birds) have the ultraviolet-sensitive cone and nonpasserines the violet-sensitive cone. Parrots are one of a number of exceptions among the nonpasserines, possessing the ultraviolet-sensitive cone. This fourth cone not only allows birds to see in the UV, but also adds an extra dimension to their color vision system. Four-dimensional color vision means birds are capable of seeing colors beyond human comprehension.

What are the benefits of UV vision?

The possession of UV vision allows parrots to signal each other via UV reflecting plumage. Most parrots exhibit some form of UV reflectance, typically associated with brightly colored feather patches used in courtship. The Budgerigar *(Melopsittacus undulatus)* has been used to explore the role of UV reflectance in mate selection. Sophie Pearn and colleagues from the University of Bristol tested female preferences for males in the presence and absence of UV light. They found that females consistently preferred UV reflecting males, indicating that UV reflectance plays a role in sexual signaling. Conversely, same sex interactions were not influenced by UV reflectance. It's sometimes argued that birds are unable to see distinct UV colors, and that UV reflectance serves only to intensify colors. The researchers were able to demonstrate that female preferences were based on the contribution made by UV reflectance to plumage color, as opposed to any effect on plumage brightness.

Parrots use UV signals as a secret communication channel, accessible only to species with the appropriate system of color vision. This scenario represents a neat solution to a conundrum faced by the males of many species—how to attract a mate while avoiding the attention of predators. Raptors are the principal predator of free-flying parrots. Their violet-sensitive cones are less sensitive than the ultraviolet-sensitive cones of their prey. Studying Eclectus Parrots *(Eclectus roratus)* on Cape York, Rob Heinsohn and colleagues demonstrated that males exploit these differences in UV sensitivity to reduce their exposure to predators, while maintaining their ability to compete for females at nest sites. UV reflecting plumage enhances the visibility of male Eclectus Parrots when viewed by competitors against the trunks of nest trees, without compromising the camouflage value of their green plumage. This study, which combined traditional field biology, modern technology, and a contemporary understanding of avian color perception, provides a template for future exploration of the nature and function of parrot coloration.

Do glowing parrots really exist?

Many parrots have feathers that fluoresce when viewed by humans in darkness under intense UV "black" lamps. This "glowing" plumage is caused by yellow pigments that absorb UV wavelengths and reemit them at longer wavelengths in the human-visible spectrum. This phenomenon gained widespread publicity following the publication in 1991 of photos accompanying an article by Walter Boles, who wondered if fluorescence served as a signal or was a biologically insignificant property of yellow pigments. An indirect signaling role for fluorescence is suggested by the location in Budgerigars *(Melopsittacus undulatus)* of UV absorbing fluorescent feathers adjacent to the UV reflecting violet-blue cheek patch, resulting in high levels of contrast and potentially greater signal strength. Franziska Hausmann and coworkers found colocation of UV reflective and fluorescent plumage to be widespread among Australian parrots.

Kathryn Arnold and colleagues undertook an innovative experiment to test for evidence of fluorescent signaling in Budgerigars. Sunblock was applied to the fluorescent crown and cheek feathers of males and females to reduce the amount of UV light reaching these areas and thus the degree of fluorescent reemission. Birds treated in this way were shunned by the opposite sex, suggesting fluorescence functioned as a sexual signal. In a separate experiment on Budgerigars, a team of scientists from the University of Bristol reached a different conclusion. They used combinations of UV filters to manipulate fluorescence, concluding the phenomenon had no impact on male attractiveness. Different methodologies may have been responsible for the contradictory findings of the two studies.

The application of sunblock to a feather patch will reduce not only its fluorescence, but also its UV reflectance. As UV reflectance is an important factor in mate choice, the University of Bristol team felt this may have unintentionally influenced the results obtained in the sunblock study. Kathryn Arnold argues the potential for such bias was small given the low levels of UV reflectance recorded from the crowns of the first generation wild-type Budgerigars used in her experiment. It is possible the filters employed by the University of Bristol scientists may have affected their results. By filtering out the UV light needed to cause fluorescence, they also filtered out the UV light needed to cause UV reflectance. The latter occurred over the whole body, which may have made birds appear odd to other Budgerigars.

Is there a reason for the color patterns of parrots?

The color displays of parrots are used to convey information about the quality of an individual. Juan F. Masello and colleagues have undertaken a long-term study of the breeding biology of Burrowing Parrots *(Cyanoliseus patagonus)* at El Cóndor on the Atlantic coast in central Argentina. This is the largest known colony of the species, numbering more than 30,000 active nests in some years. This parrot is olive-brown with a distinctive red abdominal patch shared by males and females. Researchers measured the size and intensity of the red abdominal patch and compared these to other traits that might reflect an individual's quality. They found that abdominal patches were larger in years when food was abundant, and a positive relationship existed between male patch size and body condition. The intensity of color was positively related to body condition in females and body size in males. These findings suggest that red abdominal patches in Burrowing Parrots are a signal of individual quality. This is borne out by the fact that the nestlings of males with bigger and more intense patches grow faster and weigh more than those of less ornamented birds.

It is in a female's best interests to choose a male that is in good condition as he is more likely to have access to the resources required for breeding and less likely to harbor disease or parasites. Importantly, he will contribute his good genes to her offspring. Among birds generally, there are numerous examples of choosy females selecting mates based on pigment or structural coloration. The system relies on color displays providing a reliable indication of male condition. This honesty is achieved via the direct (e.g., energy expenditure) or indirect (e.g., increased predation risk) costs arising from these displays. Variation in individual condition is expressed through the size, quality, or symmetry of the color patch. It is accepted that sexual selection is a major driving force in the evolution of color signals in birds, though it has been demonstrated for only one parrot species in the wild.

As part of their study of Burrowing Parrots at El Cóndor, Juan F. Masello and colleagues characterized the red abdominal patch in males and females from around 40 pairs. They found a correlation between abdominal patch sizes within pairs. Males with a large abdominal patch were likely to be paired with females having a large abdominal patch. This pattern of males and females apparently sorting themselves into pairs based on similarities in condition, as expressed through color patterns, is termed assortative mating. It is thought to be a result of mate choice by males and females, though other explanations not involving sexual selection are possible. If color patterns are age related then apparent examples of assortative mating could be explained by males and females forming a bond at a young age and growing old together.

Color patterns may facilitate the learned recognition of individuals. Humans are capable of recognizing individual parrots, especially among species with a high degree of variability in the size and distribution of

Juan F. Masello and colleagues studied the breeding biology of Burrowing Parrots *(Cyanoliseus patagonus)* at El Cóndor on the Atlantic coast of central Argentina.

color patches. John Pepper, studying Glossy Cockatoos *(Calyptorhynchus lathami)* on Kangaroo Island in South Australia, was able to recognize each individual in a flock of around 30 birds through natural markings. Pepper identified a dominance hierarchy within the flock. This hierarchy was stable throughout the year and few fights were recorded. The cockatoos, like the researchers, were able to recognize individuals and knew the position of each bird in the hierarchy relative to themselves, and probably to other birds as well. This reduced conflict and maximized the time available for activities such as foraging. The extent to which plumage variability assisted birds in identifying individuals is not known, but it likely played a role along with other traits (e.g., behavior).

The color of feathers can influence their resistance to bacterial degradation. A study led by Edward Burtt recently tested the resistance of different colored parrot feathers to feather-degrading bacteria. White and yellow feathers were found to break down faster than black, blue, green, or red feathers. Increasing concentrations of yellow or red psittacofulvins slowed the degradation rate. The researchers concluded that in addition to their signaling role, color patterns in parrots may have evolved to resist bacterial degradation.

Biologists must abseil down the cliff at El Cóndor to check the nests of Burrowing Parrots *(Cyanoliseus patagonus)*.

Fabián Llanos

Are there color differences between males and females?

Parrots are often described as being sexually monochromatic, the coloration of adult males and females being similar or differing only slightly. While this is true for many Neotropical species, sexual dichromatism is more common in parrots from other regions. Color differences between males and females are usually expressed through plumage, though differences in the color of the bill, cere, iris, and eye-ring also occur. Caution is required when deciding if a species is sexually dichromatic. Historically, these decisions were based on human perceptions of color. More recently, advances in our understanding of avian color vision have led to the realization that birds are capable of seeing a greater variety of colors than humans plus colors outside the realm of human experience. This raises the possibility that birds that appear sexually monochromatic to humans may be perceived as sexually dichromatic by other birds. Scientists have used spectrometry to seek out this "hidden" sexual dichromatism.

To the human eye, male and female adult Burrowing Parrots *(Cyanoliseus patagonus)* are similarly colored. Using spectrometry, Juan F. Masello and colleagues demonstrated color differences between the sexes in wild Burrowing Parrots. Green plumage was brighter in males than females, while the reverse was true for blue plumage. Reflectance curves obtained from the red abdominal patch of males and females also differed. While slight, these color differences were statistically significant. Blue-fronted Amazons *(Amazona aestiva)* are considered to be sexually monomorphic. Studying captive birds, Susana Santos and coauthors were able to determine the sex of individuals based on reflectance values obtained for the forehead, shoulder, and wing tip. While minor differences in reflectance values may not be biologically functional, they permit the development of models capable of predicting the sex of parrots using spectrometry. The development of a noninvasive technique that allows immediate gender determination will benefit field studies and captive breeding programs.

Among sexually dichromatic parrots, the male is usually more colorful. Using skins from Museum Victoria, Alice Taysom and colleagues examined the contribution made by pigment and structural coloration to sexual dichromatism in Australasian parrots. Measuring patch area, they found males had larger structure-based (blue) and psittacofulvin-based (yellow-red) color patches than females. A spectrometer was used to measure the light reflected from the different types of coloration. The spectral curves produced for structural blue colors showed large differences between males and females, with males possessing more conspicuous color patches. Minor differences between sexes in the spectral properties of psittacofulvin-based colors were found, but these were imperceptible to parrots. That male parrots had larger and more conspicuous color patches than females is consistent with the view that such ornamentation is a result of competition among males for access to females or the selection of males by females on the basis of their coloration. Structural blue colors appear to be especially important sexual signals in Australasian parrots. Elsewhere, Juan F. Masello has shown that the brightness of structural blue colors in Burrowing Parrots is correlated with environmental conditions.

Sexual selection acting on both males and females may explain why in many parrots both sexes are colorful. The alternative explanation is that color patterns in females are nonfunctional, but a by-product of their genetic relatedness to colorful males. Female plumage that is similar but duller than male plumage is consistent with this explanation. In many birds, predation-driven natural selection on females constrains the expression of conspicuous color patterns and drives the evolution of cryptic coloration (color patterns that make a bird difficult to see). Drab plumage makes it harder for predators to locate females sitting on open nests, reducing the likelihood of eggs or young being eaten. In this sense, natural selection is a strong contributor to sexual dichromatism. The hollow nesting habits of parrots reduce the likelihood of natural selection inhibiting the expression of conspicuous color patterns arising from genetic correlation with males. Tucked away in their nest hollow, female parrots avoid the gaze of potential nest predators. Accordingly, the nests of colorful individuals are no more likely to be predated than those of drab individuals. Greater predation pressure outside the nest hollow may be responsible for the evolution of drab coloration in females. In many species from the Australian genera *Psephotus* and *Neophema*, the colorful ornamentation of males contrasts with the lackluster plumage of females. These species typically forage on the ground, suggesting they would benefit from the camouflage afforded by duller plumage.

In a few parrots, both sexes are colorful but color patterns differ between the sexes. The most famous example of reversed sexual dichromatism is the Eclectus Parrot *(Eclectus roratus)*, where differences between the sexes are so marked they were for a long time considered separate species. Females compete strongly for nest hollows and their coloration has evolved to signal ownership of this scarce resource. Males compete for access to females in possession of a nest hollow, concentrating their attention on females who hold the highest quality hollows. Intrasexual signaling by males and females occurs in the same environment and may have resulted in similar color patterns if not for an additional selective force acting on males. Males are obliged to spend much of their time foraging for fruit in the rainforest canopy, where they have a relatively high exposure to predators. As a consequence, natural selection for cryptic coloration has moderated the impact of sexual selection. This has led to a divergence in the color patterns of males and females.

Other parrots exhibiting reversed sexual dichromatism include Rüppell's Parrot *(Poicephalus rueppellii)* and a number of cockatoos. Female Glossy Cockatoos *(Calyptorhynchus lathami)* have an irregular pattern of yellow blotches on the head and neck, while males lack color in this body region. The brightly colored head of females is thought to significantly increase mortality via predation, indicating the selective forces maintaining this color pattern are strong. While there is an age-related component to head color, there remains wide variation within the adult population. The functional role of head color in female Glossy Cockatoos has not been studied. Inland populations of Glossy Cockatoos are known to be limited by food resources, with the number of pairs attempting to breed closely linked to food availability. Competition for food resources may have played a role in the evolution of head color in female Glossy Cockatoos. Correlative studies of head color, dominance, and reproductive success may shed light on the evolution of color patterns in this species.

Do a parrot's colors change as they age?

Yes, juvenile parrots can usually be distinguished from adults on the basis of color. Among sexually dimorphic parrots, the juvenile plumage of both sexes is similar to adult females. Their sexual immaturity means juvenile males would derive no benefit from ornamental plumage displays. Accordingly, it makes no sense to invest in color signals that are costly to produce and may increase predation risk. Adopting the plumage of adult females may assist juvenile males in avoiding aggressive interactions with adult males, either by signaling their nonreproductive status or through female mimicry. The juveniles of sexually monomorphic parrots often lack distinctive adults-only color patches or possess other morphological traits that allow them to be readily identified. Juvenile Red-fronted Parrots *(Poicephalus gulielmi)* lack the orange-red highlights of mature birds, while young Brown-headed Parrots *(P. cryptoxanthus)* lack the yellow iris of their parents. Where the plumage of different age classes is superficially identical, juveniles can often be separated from adults on the basis of their duller plumage. Less commonly, juvenile plumage differs markedly from that of adults. Juveniles from the nominate Crimson Rosella *(Platycercus elegans)* subspecies are mostly bright olive-green, compared to the red-and-blue plumaged adults.

In some parrots, adult plumage is not attained for a number of years. Glossy Cockatoos *(Calyptorhynchus lathami)* are an example of such delayed plumage maturation, with males gaining their adult plumage at three to four years of age. Glossy Cockatoos feed almost exclusively on the seeds of sheoaks *(Allocasuarina* and *Casuarina* spp.). Releasing the tiny seeds from their encapsulating cones is an involved process, and it takes young birds a number of years to become proficient. Immature males are probably unable to achieve feeding rates necessary to support breeding. Regular drought means food is often in short supply, and there is likely to be strong competition for this resource. Populations have a biased sex ratio, with the more numerous males competing for access to females. Given these facts, immature males are not serious competitors for female attention. A study on Kangaroo Island, South Australia, found that adult male aggression was mostly directed toward other adult males. By signaling their low competitive ability, immature males avoid aggressive interactions and may gain access to resources from which they might otherwise be excluded. This may be critical during periods of food shortage when large flocks form to exploit a diminishing resource.

Is there much geographic variation in the color of a parrot species?

Many parrot species are comprised of subspecies distinguished on the basis of plumage color. Recognition of two or three subspecies is common, though in Rainbow Lorikeets *(Trichoglossus haematodus)* 20 subspecies have been identified. Taxonomists often differ over the number of subspecies within a species. Development of subspecies usually requires interruption of gene flow between populations, which commonly occurs when populations become geographically isolated. Geographic barriers that may isolate parrot populations and lead to the formation of subspecies include oceans, rivers, and mountains. Isolated populations evolve over time, resulting in the development of well-marked morphological differences. These differences may be the result of populations adapting to different environmental conditions, but sexual selection is likely to be an important driver of color pattern variation in parrot subspecies. This occurs when female preferences for male coloration differ between populations or when mutations in male plumage are preferentially selected by females. Sexual signals are somewhat arbitrary, and an orange crown may serve as well as a red crown in signaling male condition.

The Orange-cheeked Parrot *(Pyrilia barrabandi)* is distributed across the western half of the Amazon Basin. It occurs as two subspecies, a yellow-cheeked form north of the Amazon River and an orange-cheeked form south of the Amazon River. Molecular studies undertaken by Jessica Eberhard have shown these two taxa diverged less than one million years ago, long after the formation of the Amazon River. Interruption of gene flow could have occurred in a number of ways. Major climatic and vegetation changes during the Pleistocene (1.8 million years ago to 10,000 years ago) may have isolated populations of the parent species in refugia on either side of the Amazon River. They became differentiated in these climatic refuges, before subsequently expanding their range and coming into contact along the Amazon River. Even in the absence of Pleistocene refugia, the Amazon River may inhibit movement sufficiently to allow differentiation of populations to the north and south. Jessica Eberhard notes that any divergence in behavior between populations (e.g., vocalizations) would further serve to isolate them by limiting the reproductive potential of birds that occasionally cross the river. The relative role of climate and riverine barriers in the evolution of Amazon Basin taxa remains controversial, though the two hypotheses are not mutually exclusive.

Parrot species that occur over large geographical areas can exhibit gradual variation in plumage color across their range. The mechanisms responsible for such geographical variation are similar to those leading to the divergence of island populations, though the interruption of gene flow is a function of distance and dispersal behavior rather than any physical barrier. Because gene flow is slowed rather than halted, there is a reduced likelihood of populations diverging sufficiently to be considered subspecies. Adjoining populations will be genetically similar due to the frequent interchange of individuals, while populations occurring on either side of the species' range will be genetically distinct due to diminished gene flow across the intervening populations. Populations at either end of a species' range (terminal populations) may be sufficiently different to behave as separate species if they were to ever meet. This has led scientists to search for situations where terminal populations are brought into contact due to the doughnut-like nature of a species' range. Only a small number of these "ring species" have been identified.

The Crimson Rosella *(Platycercus elegans)* is found in southeastern Australia, with a number of isolated populations in northeast Queensland. Several subspecies have been recognized, which can be grouped into three broad color forms. Crimson forms occur along the coast and inland as far as the western slopes of the Great Dividing Range; Yellow forms are found along major inland river systems that have their headwaters in the Great Dividing Range; Adelaide forms occupy the coast and ranges in the vicinity of Adelaide. The plumage of Adelaide forms is intermediate between Crimson and Yellow forms and is itself highly variable, ranging from orange-red to yellow-red. The distribution of the species is ring-like in southeastern Australia, encircling a band of unsuitable semiarid habitat. It has long been suggested that Crimson Rosellas are a rare example of a ring species, with terminal populations of Yellow and Crimson forms meeting on the western slopes and connected via Adelaide-form populations half way round the ring. This view has informed the decisions of some taxonomists to treat the complex as three distinct species.

Leo Joseph and colleagues tested the Crimson Rosella ring species hypothesis by using molecular data to look for a genetic break between terminal populations of Crimson and Yellow forms on the western slopes. While they found a genetic break at the expected location, discontinuities at other locations were inconsistent with the ring species hypothesis, which predicts a gradual change in the genetic makeup of individuals around the ring. Three separate genetic groups were identified, but these did not coincide with the distribution of the three color forms. This suggests that color forms are not due to the past history of populations, but are likely to have evolved in response to existing environmental factors. Crimson forms occur in the relatively moist, closed forests of the coast and ranges, while

The St. Vincent Amazon *(Amazona guildingii)* is one of the few polymorphic parrots, having yellow-brown and green color morphs. Heinz Lambert

Yellow forms are found in drier, more open forests that border inland river systems. These quite different environments may favor the selection of markedly different color patterns. The Adelaide form is especially interesting in this regard, as its distribution coincides with a north-south environmental gradient. Birds are orange-red in the wetter southern part of the range, becoming increasingly yellow in drier northern areas.

In a few parrots, discrete categories of plumage color occur within rather than between populations. Such polymorphism is of interest because colorful pigments and perhaps structural colors may be involved. This contrasts with other bird orders, for which polymorphisms due to differences in melanin distribution are the norm. The existence of light and dark morphs in many bird species are examples of melanin-based polymorphisms. The St. Vincent Amazon *(Amazona guildingii)* is perhaps the best known of the polymorphic parrots, having yellow-brown and green color morphs. Yellow-brown morphs predominate, comprising around 85% of the population by one estimate. The mechanisms responsible for the maintenance of polymorphisms are poorly understood. It is possible that individual color morphs are favored under different environmental conditions. The preponderance of green morph St. Vincent Amazons on the eastern side of the island suggests that birds in that area derive some benefit from possessing green plumage. Alternatively, the preferential selection of green birds as mates may be sufficient to maintain this morph. Other polymorphic parrots include the Dusky Lory *(Pseudeos fuscata)*, whose color patches can be yellow or orange, and the Papuan Lorikeet *(Charmosyna papou)*, whose red body plumage is replaced by black in a melanistic morph that occurs at higher altitudes. The endangered Malherbe's Parakeet *(Cyanoramphus malherbi)* of New Zealand was thought to be a color morph of the more widespread Yellow-crowned Parakeet *(C. auriceps)*, but is now considered a distinct species.

Chapter 4

Parrot Behavior

Are parrots social?

I pulled off the quiet country road to examine a bundle of green feathers lying on the verge. As I got out of the car, a flock of Superb Parrots *(Polytelis swainsonii)* exploded from the lush weeds growing by the road. The flock landed in a nearby tree, allowing me to quickly count 25 birds. The sexes were easily distinguishable, males having a distinctive yellow forehead and a yellow throat underlined by a red crescent. The roadside corpse turned out to be a female, less colorful than the male but still a beautiful bird with her orange bill and blue-grey cheeks. She had been feeding on wild oats, trampling the tall green stems to access the maturing seed. Disturbed by an approaching vehicle, she had flown across the road and been struck. The flock had looked on quietly while I played crime scene investigator, but took off calling loudly when I started my vehicle. They headed west, while I continued east. I hadn't traveled far before coming across another flock of Superb Parrots, this time feeding in a flowering Yellow Box *(Eucalyptus melliodora).* Sixteen birds were distributed across the outer branches, some hanging upside down to reach the blossoms. As I counted and sexed this flock, a group of ten birds roared past on the opposite side of the road, grating calls heralding their approach.

Experience had taught me roadsides were a good option for finding Superb Parrots, providing important foraging habitat and pathways in a landscape largely devoid of native vegetation. Nevertheless, observing three flocks of this threatened species in quick succession made it something of a red-letter day. The noisy spectacle they presented emphasized the social nature of the species. Differences in the size and behavior of flocks sug-

The town of Boorowa in country New South Wales has adopted the Superb Parrot *(Polytelis swainsonii)* as its emblem. It promotes itself with the slogan "Superb Parrot, Superb Country." Matt Cameron

gested each had a unique identity, hinting at the complex nature of relationships within the local population. Scientists have long been fascinated by the way animal groups are organized. Much of this research has focused on mammals having obviously complex social relationships. Parrots have received less attention, the collection of data hampered by an inability to identify and maintain contact with individuals that are often hidden from view and capable of making rapid long-distance movements. The few detailed studies on social organization have relied on captive populations or been undertaken at locations where wild birds congregate and can be easily observed. Exceptions include a handful of studies on Australian cockatoos and Costa Rican parrots.

The late Ian Rowley (1926–2009) was responsible for some of the earliest studies on the behavioral ecology of parrots. One of Australia's most decorated ornithologists, he is best known for his work on fairy-wrens, corvids, and cockatoos. His cockatoo research was undertaken in the Western Australian wheatbelt and made use of wing-tags to facilitate the recognition of individuals. The impetus for his eight-year study of Galahs *(Eolophus roseicapillus)* was their status as agricultural pests, while a subsequent seven-year study of Pink Cockatoos *(Lophochroa leadbeateri)* was motivated by concerns about the species' conservation status. Graeme Chapman, a respected Australian natural history photographer, assisted with the latter study. These two studies were part of a larger program of research on the ecology of endemic cockatoos in Western Australia being undertaken by the Commonwealth Scientific and Industrial Research Organization (CSIRO). The CSIRO was charged with undertaking research on species of economic importance, though this was broadly interpreted by staff and management. The way Rowley came to work on the Galah has entered Australian ornithological folklore. With the completion of his work on corvids in eastern Australia, he was told by Harry Frith (a famous naturalist and then Chief of the Division of Wildlife Research) that "if you want to continue working on birds you will have to go to Western Australia and work on the Galah," which was "the last available pest species."

Rowley found that Galahs nesting in woodland remnants formed stable associations that traveled together between roost sites and foraging areas, often joining up with other groups en route. The use of assembly areas provided an opportunity for groups to coordinate their activities. Juvenile and immature Galahs formed large flocks that ranged widely across the landscape in search of food. These nomadic flocks also contained adults that had not yet paired or had lost a partner. The situation was somewhat different for Pink Cockatoos. Their nests were widely spaced and there was limited opportunity for pairs to form groups. Instead, they associated with small, locally stable groups of immature birds that were active in the vicinity of their nest. These groups were part of local flocks of around 50 birds, which were more cohesive outside the breeding season. Membership of local flocks was relatively stable, though individuals were recorded moving between flocks. Local flocks sometimes coalesced into groups containing as many as 250 birds.

Glossy Cockatoos *(Calyptorhynchus lathami)* possess a number of characteristics that facilitate behavioral studies. Firstly, variable plumage allows for the recognition of individuals in well-studied flocks. Secondly, they feed for long periods at a single location and may return repeatedly to preferred sites. Finally, individuals are very approachable. Taking advantage of these attributes, John Pepper undertook a study of the behavioral ecology of the Glossy Cockatoo population inhabiting Kangaroo Island, South Australia. The work was part of his PhD, undertaken through the University of Michigan. Pepper found relatively stable participation in a flock of around 30 individuals, one of a number in a population of approximately 150 birds. All members of the flock associated with each other, though they usually formed small groups comprising pairs or families. There were consistent associations between particular pairs, though these relationships could change between years. It was possible that pairs benefited from these associations through the cooperative defense of food resources. Generic relatedness is another explanation, though maintenance of associations between years would be expected in this case.

Scientists from the Cornell Laboratory of Ornithology have been investigating the part played by vocal learning in the social lives of parrots. This research has been ongoing since the early 1990s, much of it undertaken in the Area de Conservación Guanacaste of northwestern Costa Rica. This area has proven an ideal research site, parrots being abundant and relatively easy to monitor. Vocal exchanges between Orange-fronted Parakeets

Pair of Galahs *(Eolophus roseicapillus)* at their nest hollow. Heinz Lambert

(Aratinga canicularis) have been a particular focus in recent years, researchers investigating the response of wild flocks and short-term captive flocks to recorded contact calls. The results of this research have led Jack Bradbury to conclude that parrot societies fit the fission-fusion model applied to social mammals as diverse as chimpanzees, dolphins, and elephants. In communities of these species, groups form and dissolve in ways that best serve the interests of group members. Membership of groups is often unpredictable, though sex, age, and reproductive status provide an overarching framework. Fission-fusion societies are socially complex because of the number and variety of interactions that are possible. They place a premium on mutual recognition and the ability to quickly mediate interactions between unfamiliar individuals. There is a growing body of research on the importance of vocal communication in facilitating complex social networks in parrots. Fission-fusion societies are often linked to the evolution of cognitive ability because of the advantage gained by individuals able to assess the value of particular associations and capable of planning beneficial social interactions.

Kangaroo Island is home to the Kangaroo Island Glossy Cockatoo *(Calyptorhynchus lathami halmaturinus)*. Michael Barth

Why do parrots form flocks?

Flocking helps parrots avoid predators. A six-year study of breeding Peregrine Falcon *(Falco peregrinus)* diets in Canberra, Australia's capital, found that parrots made up more than half the diet by weight in each year of the study. Galahs *(Eolophus roseicapillus)* were a particularly important prey item, a finding mirrored in other Australian studies. Their use of open habitats and relatively large size make Galahs an ideal target for Peregrine Falcons. The Canberra study found the proportion of Galahs in the diet was correlated with the number of Galah flocks, not with the total number of Galahs or the size of flocks. Penny Olsen and her coauthors made the insightful observation that each flock of Galahs provides a Peregrine Falcon with a single chance to kill a bird. In effect, a flock represents a single prey item. With many birds keeping an eye out, the likelihood of a predator approaching a flock undetected is reduced. Raptors may be confused by a wheeling, compact flock of parrots and be unable to isolate a single bird for attack.

Participation in flocks allows birds to spend less time scanning for predators without increasing predation risk, making more time available for foraging. David Westcott and Andrew Cockburn examined vigilance behavior of Red-rumped Parrots *(Psephotus haematonotus)* and Galahs in Canberra. They found that as flock size increased, the vigilance behavior of individuals within the flock decreased. Birds feeding alone or in pairs spent 50% of their time on vigilance, while birds in flocks of ten or more birds spent as little as 10% of their time on this activity. This reduction was possible because the probability of at least one individual being vigilant increased with flock size. Thus, a reduction in individual vigilance was matched by an increase in flock vigilance. They also found that birds in the center of the flock were less vigilant than those on the edge, suggesting that position within the flock was an important influence on predation risk. A study of naturalized Monk Parakeets *(Myiopsitta monachus)* in Chicago similarly showed that flock membership reduced the time individuals spent keeping an eye out for predators.

Flock size varies with food abundance, suggesting the benefits of flocking extend beyond predator avoidance. I studied flock size in Glossy Cockatoos *(Calyptorhynchus lathami)* over a period when food resources varied substantially. I found flock size was significantly greater during periods when food was reduced, a pattern evident in other cockatoos. Mechanisms of flock formation at these times and the benefits derived from participa-

In the winter months, Galahs *(Eolophus roseicapillus)* can often be observed feeding on suburban lawns. Matt Cameron

tion in such flocks are not well understood. Birds may be actively recruited into flocks, follow successful foragers from roost sites, or simply join birds they observe feeding. Flock members presumably benefit from improved access to food resources in the longer term, though food intake may be reduced in the short term. Variation in flock size over time suggests that birds are constantly assessing the costs and benefits of flock participation.

If food availability is a factor in flock formation, we would expect parrots from arid regions to form larger flocks than parrots occupying wetter, more productive environments. In one of the earliest behavioral studies of small Australian parrots, Christine Cannon contrasted flock size in Eastern Rosellas *(Platycercus eximius)* occupying a dry site with that of Pale-headed Rosellas *(P. adscitus)* occupying a wet site. Eastern Rosellas formed larger flocks and exhibited greater variability in flock size than Pale-headed Rosellas, which was attributed in part to the less predictable nature of the Eastern Rosellas food supply. Studying flock size in parrots occupying lowland tropical forest, James Gilardi and Charles Munn were struck by the fact that across a wide range of species the average flock size was only two to four individuals. They raised the possibility that forest-dwelling parrots may be less social than their open-country cousins, with the arid nature of habitats occupied by the latter species a possible explanation. A complicating factor is that open-country species may form large flocks due to their greater exposure to predators. Predator avoidance and foraging efficiency are not mutually exclusive advantages, and flock formation is likely to be a result of a number of factors.

Flocks may form to allow the efficient exploitation of abundant food resources. In southeastern Brazil, researchers found that flocks of Maroon-bellied Parakeets *(Pyrrhura frontalis)* were largest during the dry season. At this time, birds were feeding on palm fruit that was abundant and widespread throughout the study area. Parrots may gather to forage on exotic foods, especially during periods of natural food shortage. Pine plantations surrounding Perth provide an important food source for Carnaby's Cockatoos *(Calyptorhynchus latirostris)* in the nonbreeding season. In the 1940s, foresters reported flocks of 5,000 to 6,000 birds feeding in these plantations. Despite a decline in the overall size of the cockatoo population, it is still possible to observe flocks containing thousands of birds. In the Kimberley of Western Australia, one study found that a flock of Little Corellas *(Cacatua sanguinea)* containing tens of thousands of birds formed to take advantage of sorghum crops during periods when native grass seed was unavailable. The exploitation of exotic foods by large, mobile flocks of parrots gives a misleading impression of abundance. Parrots that gather on crops can do considerable damage and are vulnerable to persecution.

Birds benefit socially from participation in flocks. In Galahs, membership of the local flock allows young males and females to develop associations that lead to the formation of mated pairs. These flocks also function as lonely-hearts clubs for adult birds that have lost partners. Ian Rowley noted that paired males often associated with unpaired females from the local flock. He concluded that males were probably aware of potential partners in the event they lost their mate. This ensured that widowed or divorced birds quickly formed new pairs, which reduced the likelihood of them missing a breeding season. Flocks also provide an opportunity for young birds to learn the ropes from older, more experienced flock members. This is especially true for species that have limited parental care or take several years to reach maturity. The knowledge held by older flock members on the location of food and water is especially valuable in environments where these resources are patchily distributed in time and space.

Age-related differences in flock size provide additional indirect evidence for the benefits of flocking. Immature parrots and adults separate into distinct flocks in some parrots, with immature flocks typically larger than those of adults. This is not surprising given immature birds are likely to benefit disproportionately from the predation, feeding, and social advantages conferred by flocking. Robert Magrath and Alan Lill studied the behavior of adult and immature Crimson Rosellas *(Platycercus elegans)* inhabiting wet sclerophyll forests east of Melbourne, Victoria. The two age classes can be distinguished on the basis of color, the red-and-blue plumage of adults contrasting with the mostly green plumage of immature birds. They found that immature flocks were typically larger than those of adult or mixed age flocks, probably due to immature birds being restricted to inferior habitats where resources were less predictable and predation risks possibly greater. Because they needed to forage for longer periods to meet their energy requirements, immature birds would have particularly benefited from behaviors that increased the time available for feeding.

How do parrots spend their day?

Parrots spend most of the day resting or feeding, interspersed with bouts of socializing. The pattern of daily activity is consistent between species. Birds rouse from their night roost at dawn, preening and moving around the roost tree. They often make short flights and interact with birds that have roosted nearby. After this early morning socialization, they depart for feeding areas. These may be nearby or involve a flight of many kilometers. Birds feed until their crop (an enlargement in the esophagus used to store food) is full and then rest in the canopy while this food is digested. They may feed again before taking an extended break at midday. Rest periods are a time for preening and social interaction. In the afternoon, birds feed for a short period before drinking and returning to their night roost. This daily routine is driven by a bird's hunger on awakening and the need to fill its crop ahead of the night fast. It results in a characteristic bimodal pattern of foraging activity, with an extended peak in the morning and a shorter peak in the late afternoon.

Social interactions are an important component of a bird's daily activities. The period between waking and departing for feeding areas is particularly busy. Puerto Rican Amazons *(Amazona vittata)* spend the first hour of each day making short noisy flights from one emergent canopy tree to another. While perched, pairs perform duets and respond to the calls of other pairs. A less vigorous period of calling occurs in the evening prior to birds going to roost. A similar pattern is evident in the Yellow-naped Amazon *(A. auropalliata)*, pairs using habitual calling locations from which they broadcast several duets before moving on. Some species make use of assembly areas or activity centers. These are often located close to food and water and characterized by the presence of large old trees. Typically, birds fly to these sites from night roosts. They are a focus for social activity, birds calling loudly and making frequent short circling flights. Birds may feed and drink at activity centers before coalescing into larger groups and dispersing to other feeding sites.

Water is not a concern for parrots inhabiting wet tropical environments. Moisture is widely available and fleshy foods mean birds are less reliant on standing water. Granivorous species occupying dry environments need to drink at least once a day. Water points are widely spaced and birds congregate at these locations at set times. Stock troughs and farm dams provide a reliable source of water. Australian cockatoos typically drink in the evening, though when temperatures are high they may also drink at midmorning. Smaller parrots will often drink morning and evening.

The behavior of parrots at water points indicates they are vulnerable to predators at these locations. Glossy Cockatoos *(Calyptorhynchus lathami)* prefer to drink from pools within vegetated drainage lines, though will drink from open dams when no alternative exists. They fly into water points in a steady stream 30 to 60 minutes before sunset, congregating in the surrounding vegetation. They progressively descend through the canopy until a lone bird drops to the water's edge. The rest of the flock follows en masse. Drinking birds do not linger, quickly retreating to the safety of the surrounding vegetation once they have quenched their thirst. Juveniles appear naive to the predation risk and are often left stranded at the water's edge when the drinking flock flushes.

Parrots may have to travel long distances between roost sites, foraging areas, and water points. They often use fixed routes to move between these resources, providing scientists with an opportunity to understand patterns

Regent Parrots *(Polytelis anthopeplus)* drinking from a stock trough, an important source of water for many inland Australian parrots. Matt Cameron

of resource use and census populations. Flyways are obvious in fragmented or patchy landscapes, birds forced to move along strips of vegetation or cross open areas. Studying Grey Parrots *(Psittacus erithacus)* in Uganda, John Amuno and colleagues found birds moved between forest roosts and surrounding feeding areas along regular flyways. Birds preferred flyways that provided opportunities for foraging, resting, and socializing en route to feeding areas.

Olaf Wirminghaus and coworkers noted that Cape Parrots *(Poicephalus robustus)* traveled 10 to 20 kilometers between forest patches along regular flight paths. These movements were necessitated by the patchiness of the food supply and the limited availability of surface water at certain times of the year. No single forest patch was capable of supplying all the resources needed by the birds. Olaf Wirminghaus' research on Cape Parrots was undertaken as part of his PhD through the University of Natal (now University of KwaZulu-Natal). This research was cut short by his untimely death in 1996. His wife, Colleen Downs, and colleagues, especially Craig Symes, ensured the work was completed and published. These papers have made a significant contribution to Cape Parrot conservation.

The wariness displayed by some parrots, in particular smaller species,

The risk of predation when drinking means that parrots, such as these Black-cheeked Lovebirds *(Agapornis nigrigenis)*, often descend to the waters edge en masse. Heinz Lambert

when entering night roosts makes it difficult to study this aspect of behavior. Mated pairs usually roost alongside each other. Pairs may separate during the breeding season, one member spending the night in the nest hollow while the other roosts nearby. I found that male Glossy Cockatoos roosted a few hundred meters from the nest tree, returning to the same area each night. A number of species retain an association with their nest site throughout the year, including roosting in the general vicinity. Noel Snyder and coauthors reported that outside the breeding season Puerto Rican Amazons roosted in the same valleys used for nesting, with birds occasionally roosting near their nest. Similarly, pairs of Galahs *(Eolophus roseicapillus)* return to their nest site each evening outside the breeding season. This contrasts with the behavior of unpaired and immature Galahs, who roost close to the area where they have been foraging.

Some parrots form large communal roosts at sites that have been used for many decades. Some of the best-studied communal roosts are those of Amazon parrots. Timothy Wright mapped night roosts of the Yellow-naped Amazon in northwestern Costa Rica. These roosts are widely dispersed and each contains 20 to 300 birds, with individual birds exhibiting fidelity to particular sites. One roost is known to have been occupied for at least 30 years. The size of parrot roosts can vary throughout the year depending on the presence or absence of breeding birds and their offspring. Monitoring of roosts has been used to determine population size and check for long-term changes in abundance. Seasonal variation in numbers and composition over the course of the year may provide information on the proportion of pairs that breed and the number of young they produce. The traditional nature of some roosts makes their occupants vulnerable to persecution or exploitation.

How do parrots communicate?

Parrots use a variety of calls to communicate, often in combination with physical displays. Most species have 10–15 distinct calls, each having a particular function. The best known of these is the contact call, used to establish a vocal link between birds and coordinate flock movements. It also identifies individuals. A specific preflight call provides advance notice of the caller's intention to depart a location, preventing birds being left behind. Some calls vary in meaning depending on context. Ian Rowley described a screech call in Galahs *(Eolophus roseicapillus)* that was used when birds were frightened, in pain, or threatening other birds. In some species, there are specific calls associated with courtship and copulation. An example is the "kwee-chuck" call of displaying male Glossy Cockatoos *(Calyptorhynchus lathami)*. Young parrots have a specific begging call, as do females in species that practice courtship feeding.

Individual Recognition. Visitors to the Western Australian wheatbelt in spring have a good chance of encountering the endangered Carnaby's Cockatoo *(Calyptorhynchus latirostris)*, which nests in remnants of native vegetation along roadsides and railway lines. Listening for the drawn out contact call uttered by birds in flight is the best way to locate flocks. Thanks to the work of Denis Saunders, we know that Carnaby's Cockatoos can identify individuals based on this call. His work on vocal communication in Carnaby's Cockatoos was part of a larger study on their foraging ecology and breeding biology. The knowledge gained from this research, which spanned nearly three decades (1969–1996), now underpins conservation efforts for Carnaby's Cockatoos and other wheatbelt species. It would be nice to think Saunders' decision to study Carnaby's Cockatoos was based on a prescient understanding of the future impact broad-scale clearing for agriculture would have on the species. In fact, he admits it was based on a fear of heights. Having to decide between a study on Carnaby's Cockatoo and the closely related Baudin's Cockatoo, he chose the former given the latter was thought to nest in the tall eucalypt forests in the far southwest of Western Australia. Of course, a study on Baudin's Cockatoo would have been equally valuable given its current conservation status.

Saunders recognized 15 different types of vocalization produced by Carnaby's Cockatoos, but his interest was piqued by the loudest and most common call, which he referred to as the "wy-lah." He noted that incubating females responded to the contact calls of their partners by leaving the hollow and either flying to join them or awaiting their arrival in a nearby tree. The calls of other males were ignored. The ability of females to distinguish the calls of different males was confirmed by playback trials. During the breeding season of 1980, Saunders used a directional microphone and cassette tape recorder to record the calls of males returning to their nest hollows. The calls of a number of males, including their partners, were played to females present in their nest hollows. Females never responded to the calls of strange males, and nearly always responded to the calls of their mate. The ability of females to recognize their partner's calls meant they did not leave the hollow unnecessarily, which limited exposure of their eggs or nestlings to the elements or predators.

To examine parent-offspring recognition in the Galah, Ian Rowley swapped broods of wild Galah nestlings of various ages from one nest to another. The movement of small nestlings resulted in no discernible behavioral response from nestlings or their foster parents. Moving four- to five-week-old nestlings created some confusion. Nestlings of this age could recognize their parent's calls and did not beg when their foster parents approached the nest. The absence of begging caused consternation among the adults, which appeared oblivious to the replacement of their offspring. Nevertheless, their urge to feed the contents of their nest hollow combined with the nestling's hunger meant these transfers were successful. Swapping six-week-old nestlings was unsuccessful, with foster parents rejecting the nestlings. At this age, nestlings were coming to the hollow entrance to be fed, and this may have facilitated parent-offspring recognition. Recognition of offspring is delayed until just prior to fledging (their first flight), as parents don't need to identify their young until they leave the nest. Contrastingly, nestlings benefit from an ability to discern and respond to the calls of other birds early in life. They fall silent when other Galahs give alarm calls and avoid begging when there is little likelihood of being fed. Accordingly, they are less likely to draw the attention of auditory predators.

Ralf Wanker and colleagues from the Universität Hamburg studied vocalizations in Spectacled Parrotlets *(Forpus conspicillatus)*, a small green parrot known for its incessant twittering and chattering. Spectacled Parrotlets use different contact calls when interacting with different individuals. This is analogous to individuals having names for each member of their family.

During playback trials, each member of the family responded to its name when called. It appears that Spectacled Parrots have an abstract or mental representation of at least family members. The contact calls of individual Spectacled Parrotlets also allows categorization based on social relationships. The formation of mental representations of individuals and relationships is an important precursor to planning beneficial social interactions. The extent to which parrots understand their own motivations and those of other individuals remains to be explored.

Vocal learning. Parrots are able to modify vocalizations based on social experience. We are only beginning to understand how such vocal learning aids survival in the wild. Research has focused on the way vocalizations assist individuals to manage relationships in complex societies. Much of this research is aimed at identifying contact call convergence within pairs, flocks, or roosts—demonstrating vocal learning at each level of social organization. Studies of captive parrots have played an important role, providing researchers with an opportunity to manipulate social groupings. Technological advances in acoustic hardware and software have facilitated a growing number of studies on wild populations under natural conditions. Another potential benefit of vocal learning is the sharing of environmental information. For example, some primates have distinct calls for palatable foods and dangerous predators. Captive parrots are able to correctly label objects in their environment, though there is no evidence for this in wild populations. Studying the vocalizations of birds functioning as sentinels for foraging flocks is one possible line of inquiry.

Arla Hile and colleagues from the University of California investigated call convergence in captive Budgerigars *(Melopsittacus undulatus)* during pair bonding. They created artificial pairs by placing unfamiliar males and females together and then monitored changes in the contact call of each sex. Pairs rapidly developed a shared contact call, largely a result of males imitating the calls of females. Subsequent studies demonstrated that a male's inability to mimic a female's call was likely to result in the female engaging in extra-pair copulations. It appeared that vocal convergence in the weeks following pair formation provided females with an opportunity to assess a male's fitness. Failure to converge on the female's call indicates the male is of lesser quality. Changing mates in this situation may not be an option given the limited breeding opportunities afforded by the Australian outback, so females hedge their bets by seeking genetic input from other males. Natural drift in the call of female Budgerigars over time means that male fitness is constantly being challenged.

Parrot flocks may form and then split over the course of a day. Despite regular fusion and fission, membership of cockatoo flocks remains relatively stable over the longer term, suggesting birds benefit from associating with known individuals. What mechanisms do parrots employ to enable the rapid fusion and fission of flocks? Individual recognition probably plays a role, but may not be feasible or practical where large numbers of birds are involved. The vocal learning abilities of parrots allow for an alternate mechanism. When flocks of Budgerigars are created from individuals unfamiliar with each other, convergence on a common call type can occur within a few weeks. Male flocks develop a shared call more quickly than female flocks. Arla Hile and George Striedter suspect this reflects the low probability of female flocks forming in the wild, whereas male flocks are common early in the breeding season when females stay close to the nest.

A shared contact call is like wearing a football jumper that signals your membership of a particular team. If you are foraging in a large flock and suddenly a group of birds departs, it's easier to check the color of their jumpers than it is to try and recognize individuals. If the departing birds are on your team, you can quickly join them before they disappear. How do parrots benefit from maintaining stable social units? One advantage is that previously established dominance hierarchies are likely to remain in force, reducing conflict and maximizing the time available for other activities. Additionally, if birds join flocks at random they may end up covering the same ground twice, seeking out resources they had accessed the previous day. Large home ranges, lack of territoriality, and patchy distribution of resources increase the likelihood of this occurring. Stable social units may also allow for reciprocal sharing, groups acting in ways that benefit other groups in the expectation the favor will be returned.

The findings of the above studies on Spectacled Parrotlets and Budgerigars come with an important caveat—the research was conducted using captive birds. Spectacled Parrotlets employed in the Universität Hamburg studies were captive bred and sourced from the University's breeding stock or German aviculturists. Budgerigars participating in the University of California studies were obtained from a local wholesale distributor and part of a population held in captivity for many generations. It remains to be seen if wild populations of these species behave in a similar manner. Demonstrating that parrots use shared contact calls to facilitate social interaction in the wild is difficult. The Cornell Laboratory of Ornithology has been at the forefront of such research, undertaking field studies on a number of South American parrots.

Susannah Burham-Deever and colleagues studied Brown-throated Parakeets *(Aratinga pertinax)* on the island of Bonaire in the Netherlands Antilles. They observed the interaction between foraging parakeets and overflying groups, finding that foraging birds responded to the calls of overflying groups in some but not all cases. Overflying groups were more likely to settle in the general area when called to by foraging birds. Playback trials confirmed that overflying groups were responding to calls, rather than any visual signal. In this study, it appeared that foraging birds were choosing which groups they would recruit to foraging sites. Significantly, this choice appeared to be based on the contact calls of the overflying group. The researchers demonstrated that birds were capable of recognizing the calls of individual birds, but failed to find any evidence of shared contact calls within groups.

Even where group membership is relatively stable, parrots will move from one group to another. Integration into successive social groups requires individuals to learn a large number of new calls throughout their life. In addition, parrots require a mechanism for rapid mediation of interactions between unfamiliar individuals. Playback trials involving wild and captive Orange-fronted Parakeets *(A. canicularis)* found that responses tended to converge or diverge on the stimulus call. This suggests these exchanges are akin to a negotiation that may determine whether fusion occurs and hierarchies within any fused flock. Only a few individuals within flocks replied to the broadcast call, raising the possibility of birds within flocks occupying leadership positions. On the other side of the world, Judith Scarl and Jack Bradbury similarly found that Galahs responded to the broadcast of prerecorded contact calls by imitating the broadcast call. This vocal matching occurred during a single interaction that may have lasted only a few minutes.

It has long been known that each parrot has a unique contact call, which birds use to identify individuals. The capacity parrots have for vocal learning means it's possible these vocal signatures are not innate, but learned early in life. Karl Berg and colleagues investigated vocal learning in an intensively studied population of Green-rumped Parrotlets *(Forpus passerinus)* occupying a cattle ranch in Venezuela. Birds in the population are color banded and their pedigrees known. This was important, because the researchers needed to swap clutches between unrelated pairs to determine the role of genetics in call acquisition. Birds nested in artificial hollows hung on fence posts, making it easy for researchers to install audio-video equipment to record the calls of adults and nestlings. Clutches were successfully swapped among nine nests, with eight other nests serving as experimental controls. Analysis of the recordings showed the call of each adult was more similar to its mates than to other adults, indicating parrotlet pairs matched their calls. The unique calls of nestlings were similar to their siblings and to the adult birds that reared them, indicating calls were learned and not inherited. These shared contact calls no doubt help parrotlet families stay in contact once chicks have left the nest.

Geographical Variation. Researchers from Cornell University recorded contact calls of Orange-fronted Parakeets throughout their range in Costa Rica. They found that calls were relatively uniform at a scale consistent with the home range size of individuals, but varied significantly between regions. These differences were shown to be the result of gradual variation along geographic gradients. Alan Bond and Judy Diamond found a similar situation for Kea *(Nestor notabilis)* in New Zealand, where contact calls of adult and juvenile birds vary as you move along and across the Southern Alps. Geographic variation in the contact calls of Orange-fronted Parakeets and Kea may simply be an artifact of call convergence within local populations, with the calls of geographically separate populations expected to drift and thus diverge over time. Such divergence may benefit birds by preventing the inappropriate fusions of flocks. If two flocks having distant ranges converge on a temporarily abundant food resource, there is the potential for an interchange of individuals between flocks. Once the food supply is exhausted, differences in contact calls may prevent birds following the wrong flock home. Birds suddenly finding themselves in an unfamiliar environment are at a disadvantage, needing to learn from scratch the spatial and temporal distribution of food and water.

Timothy Wright has studied vocal communication in Yellow-naped Amazons *(Amazona auropalliata)* for two decades, basing himself in the dry tropical forests of northwest Costa Rica. In the first year of the project, he drove his battered Volkswagen down from the United States. A lack of funds halted his progress at the Nicaraguan–Costa Rican border, a situation resolved by the sale of his tape deck to a customs officer. The direction of his research was set when he realized Yellow-naped Amazons in the north and south of the study area had different contact calls. Occupying abandoned ranch houses within the newly declared Guanacaste National Park, Wright spent up to six months a year in Costa Rica. He mapped the parrot's night roosts, discovering the best way to locate these was to ask the locals. Contact calls of birds using each roost were recorded and compared in order to confirm the existence of dialects initially distinguished by ear. The work was extended to cover other calls in the Yellow-naped Amazon's vocal repertoire, including duets performed by mated pairs. Now based at New Mexico State University, Wright's project continues to this day, providing opportunities for a succession of undergraduate and postgraduate students.

Timothy Wright's investigations confirmed that Yellow-naped Amazons using roosts in the north had a distinct contact call, as did birds frequenting southern roosts. Roosts situated along the dialect boundary were dominated by either dialect, but a few birds at these sites were bilingual. The bilingual birds never blended dialects, always using one or the other.

The Wright Lab field team, May 2005. Timothy Wright

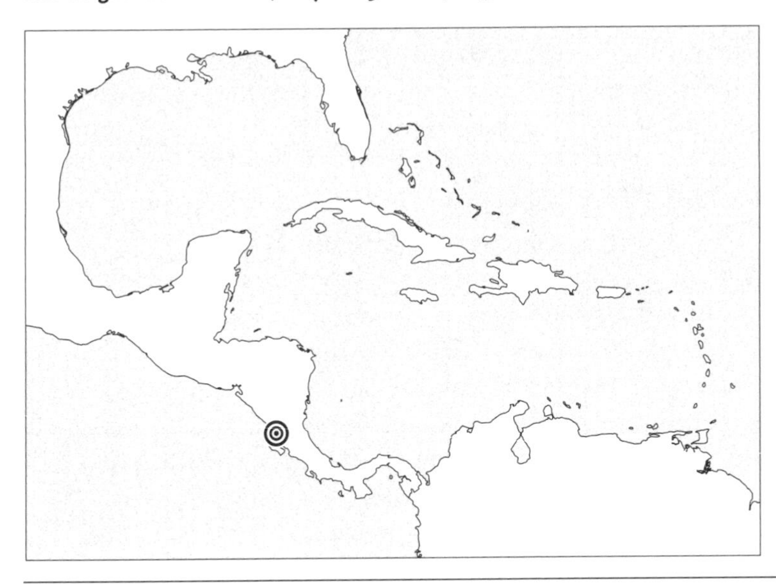

Tim Wright has spent 20 years studying vocal communication in Yellow-naped Amazons *(Amazona auropalliata)* in northwest Costa Rica.

The boundary between northern and southern dialects has remained stable over an 11-year period. Genetic studies revealed no differences between dialects, indicating that dialects were not a barrier to dispersal. Dialects appeared to be maintained by immigrant birds learning the local call type and are thus cultural phenomena. Recent translocations of birds across dialects, performed by Alejandro Salinas-Melgoza, have shown they are capable of learning new contact calls within six weeks. Nevertheless, it remains possible that reduced social fitness of immigrants due to their differing vocalizations may contribute to dialect maintenance. The research also revealed differences in contact calls between roosts within dialects. Shared calls may allow access to the roost as well as facilitate social interaction at other times. It is unclear what additional benefits accrue from the maintenance of dialects across multiple roosts. Similarly, it is difficult to explain how dialects first evolved in the absence of any genetic evidence for past geographical isolation.

How smart are parrots?

Parrots are among the smartest of all nonhuman animals. The only birds known to possess comparable intelligence are corvids. The cognitive abilities of parrots are similar in many ways to those exhibited by nonhuman primates, and in some areas approach those of young children. Assessing the intelligence of parrots in laboratory settings is not easy, but the greater challenge is deciphering how such intelligence is employed by wild populations to solve problems and enhance survival.

Cognitive Abilities. Our inability to communicate effectively with animals makes assessing their intelligence problematic. It can be difficult to distinguish innate or learned behaviors from those providing evidence of higher cognitive processes such as insight. To overcome these problems, researchers have devised behavioral tests that allow them to assess the cognitive abilities of participants. Some animals are capable of learning a code that enables communication with humans, which in turn allows a fuller exploration of an animal's cognitive ability. The vocal skills of parrots mean that communication codes based on human speech are possible. These have been most famously employed by Irene Pepperberg in her studies of the cognitive and communicative abilities of Grey Parrots *(Psittacus erithacus)*.

The ability to solve novel tasks without recourse to trial and error is considered to provide evidence of insight, which is often tested in birds by setting a string-pulling task. Here, a food reward is attached to a string which is suspended from a perch. The string and food may be encased in an open-ended plastic cylinder or otherwise shielded to prevent the food being accessed from the ground or by flight. The problem is typically solved by pulling up on the string with the bill, and clamping the string with the foot between pulls. Ludwig Huber and Gyula Gajdon set this task for young captive Kea *(Nestor notabilis)* that had no experience with string-pulling. The Kea solved the problem instantaneously without exploration. They understood the problem, developed an effective solution, and put this into practice.

String-pulling studies on Australian and Neotropical parrots have not been as conclusive as those involving Kea, though many tested individuals were similarly capable of retrieving the food reward on the first attempt. Successful completion of the string-pulling task demonstrates an appreciation for cause-effect relationships, but other factors may also be involved. One study found that Hyacinth Macaws *(Anodorhynchus hyacinthinus)* were motivated to string-pull even in the absence of food rewards. The propensity of this species for play and exploration means that string-pulling may function as its own reward. Birds may also employ some innate or learned behaviors, string-pulling having parallels with natural food harvesting. In this regard, setting counterintuitive tasks that require birds to pull down on strings to retrieve rewards may be instructive. These have been used successfully in Berndt Heinrich's studies designed to test problem solving in ravens.

The use of tools may indicate that animals understand the causal relationships between objects. There is limited evidence for tool use in parrots. One possible explanation is that parrots have little need for tools given their bill can be turned to a variety of tasks. Wild Hyacinth Macaws have been observed to wrap nuts in leaves to prevent them slipping during processing, while captive birds use pieces of wood as chocks to secure nuts in the bill. On Cape York, Palm Cockatoo *(Probosciger aterrimus)* males drum on the trunks of hollow trees using their clenched foot, a nut, or specially prepared stick. These observations have all been anecdotal. In the absence of experimental testing it is difficult to determine the extent to which individuals understood the physical processes involved. Palm Cockatoo drumming is interesting because it appears to be confined to the Cape York population. Palm Cockatoos transport prepared branches to hollows for use in the construction of nest platforms, and it wouldn't be a huge leap for one of these branches to be used as a drumstick. Birds often prepare and discard a number of drumsticks before settling on one for extended use, indicating that trial and error plays a role.

Irene Pepperberg taught a Grey Parrot named Alex to label objects using elements of English speech. Alex had labels for over fifty objects, as well as for several colors and shapes. This capacity to label objects allowed his cognitive abilities to be assessed through his answers to questions on concepts of category, same/different, relative size, absence, and number. He was presented with novel objects having both color and shape and asked, "What color?" or "What shape?" He was shown pairs of objects that shared one category but differed in another and asked, "What's same?" or "What's different?" He was shown groups of objects and asked to label subsets of these. For example, if presented with a tray holding one blue box, three green boxes, four blue cups, and six green cups, he would be asked "How many green cup?" Alex did well on these and similar tasks, typically answering around 80% of questions correctly.

Alex's numerical competency improved over time. In early tests of same/different Alex was taught to label an absence of difference as "none." He later transferred this term without training to studies of relative size. His comprehension of this concept progressed to the point where unprompted he used the term "none" to signify the absence of a subset of objects from a tray. The use of "none" in this context is somewhat analogous to "zero," an abstract concept he was thought to have figured out for himself. Similarly, Alex came to an understanding that Arabic numerals were representations of a certain number of objects and understood the relationship of one numeral to another. For example, he knew that the numeral five was "bigger" than a collection of three physically larger blocks. Comparisons between species and studies are difficult, but Alex's performance across the range of studies he participated in was variously considered comparable to that of primates, dolphins, and young children.

The long period of training received by Alex may be a factor in his understanding of numerical concepts. How does Alex compare to Grey Parrots that have not received such training? University of Paris researchers investigated the ability of untrained Grey Parrots to discriminate between discrete and continuous amounts of food. Three young Grey Parrots were presented with two sets of seeds, with each set containing between one and five seeds. Alternatively, they were presented with two amounts of parrot formula, with each amount varying between 0.2 and 1.0 milliliter. The parrots chose the larger amount for most possible combinations, though they performed best at lower ratios (e.g., one versus five seeds). This is termed the analogue magnitude mechanism, and is known to be used by humans and other animals. Alex was thought to use counting to correctly identify quantities, and in later studies demonstrated an ability to add small numbers. Parrots appear capable of using a variety of mechanisms to determine quantity, dependent on the task at hand and the level of training they receive.

Evolution. Why do parrots have special cognitive abilities? Intelligence is a function of the size, structure, and wiring of the brain. If absolute

Alex. Arlene Levin-Rowe

brain size was important, we would not expect parrots to perform as well as chimpanzees on cognitive tests. In fact, scientists have found that brain size relative to body size is the best predictor of intelligence. The relative brain sizes of parrots and chimpanzees are similar, smaller than humans but larger than most other animals. Different wiring and processing within avian brains may allow them to perform at levels comparable to those of primates despite their smaller absolute size. The brain is a costly organ for the body to run, and animals are unlikely to have evolved large brains unless they delivered greater fitness. This is especially the case for birds, where additional weight increases costs associated with flight. The evolution of a large brain has improved the capacity of parrots to cope with the social and ecological challenges they encounter on a daily basis.

The primate brain is thought to have evolved in response to the social challenges of group living. The same may be true for parrots, with natural selection favoring birds that are better able to develop and maintain social relationships. Participation in groups imparts greater fitness because of the benefits they afford in terms of predator avoidance, food finding, and mate selection. Social competition within groups may act as an additional selective pressure. An alternate hypothesis is that environmental challenges associated with foraging were responsible for the evolution of parrot cognitive abilities, with these subsequently allowing the evolution of complex societies. It cannot be said with any certainty whether social or ecological factors were responsible for the evolution of large brains in parrots; both may have played a role. The evolution of large brains is also inextricably linked to the slow life histories of parrots. Slow growth rates, delayed maturation, and long lifespans provide time for brains to develop, for birds to learn, and for benefits to accrue from this learning.

There is considerable variation in relative brain size between parrot species. To investigate the role challenging environments played in the evolution of brain size, Cynthia Schuck-Paim and colleagues analyzed the relationship between this characteristic and climatic variability in Neotropical parrots. They found that parrots presently occupying climatically unstable habitats have larger brains than species found in more stable environments. This relationship holds for comparisons within some genera. The fact that climatically unstable and stable environments are analogues for open and forest habitats makes interpreting these results difficult. Open-country parrots appear to live in more complex societies, suggesting that social factors may be responsible for the observed differences in brain size. Of course, social complexity may have evolved in response to climatic instability. Analyses of brain size versus social complexity will help resolve this issue, but this requires additional data on the behavioral ecology of parrots.

Ecological Benefits. Some parrots occupy harsh environments, characterized by limited food resources and marked variation in food availability over time. A flexible approach to food gathering is required by species resident in these habitats. An ability to switch from one food to another as the abundance of each waxes and wanes is required, as is a capacity to identify and exploit novel food items. The Kea, a New Zealand parrot renowned for its curiosity, is a classic case. It occurs in alpine areas and has been recorded feeding on over 100 species of plants and animals. The Pink Cockatoo *(Lophochroa leadbeateri)* faces similar environmental challenges to the Kea, inhabiting arid and semiarid areas of Australia. One study recorded it feeding on around 30 food items, including seeds, fruits, flowers, roots, stems, and insect larvae. The cognitive abilities of Kea are well documented, but those of Pink Cockatoos have not been studied.

Studies of captive birds have shed light on how ecology drives exploratory behavior in parrots. In several studies, Claudia Mettke-Hofmann and coauthors exposed a wide range of parrot species to novel objects in aviary settings. Compared to mainland species, island parrots were quicker to approach novel objects and explored them for longer periods. Competition for limited island resources can be intense, and parrots would benefit from behaviors that enable the early discovery and utilization of food items. The researchers also noted that species with a high proportion of fruit in the diet were highly motivated to explore. Fruit is patchily distributed in space and time. Exploration allows the gathering of information on its current and future availability, while well-developed spatial memories assist efficient exploitation of the resource. Long-term memories of fruiting pat-

Keas *(Nestor notabilis)* have an inquisitive nature, which often gets them into trouble. Matt Cameron

terns enable parrots to turn up at the right place at the right time, while short-term memories ensure that previously exploited fruit trees are not needlessly revisited.

Parrots usually need to feed on a variety of foods to survive. They must develop techniques for opening and processing these foods, many of which have evolved defenses against predators or are otherwise protected. In some species, young birds take a number of years to become proficient at food handling. Claudia Mettke-Hofmann and coauthors found that the presence of nuts in the natural diet was positively associated with exploration periods of novel items by captive parrots. Innate behavior or even trial and error may not be sufficient to solve the problem of some hard-to-open nuts, which may require the application of insight or other problem solving mechanisms.

Specialization on one or two foods has its own challenges. There is often a trade-off between security of food supply and return per unit effort, with dietary specialists having to feed for long periods to satisfy their energy requirements. In an effort to maximize food intake rates, dietary specialists are highly selective in their feeding. They make foraging decisions at multiple spatial scales, feeding in the patch with the most trees, the trees with the most fruit, and the fruit with the most seeds. Animals thought to have limited cognitive ability are capable of completing these tasks, but categorization of food resources and hierarchical decision making would benefit from high-level cognition.

Can parrots talk?

Many parrot species have a well-deserved reputation for talking in captivity, with the utterances of some parrots indistinguishable from those of their owners. Parrots are often considered mindless mimics, so much so that "parroting" has come to mean repeating another's words without understanding their meaning. Scientists distinguish between mimicry and imitation, though they debate the precise meaning of these terms. The cage-confined porch-dwelling cockatoo uttering to all and sundry, "Polly want a cracker!" is a mimic. He has heard and learnt the phrase, but repeats it endlessly without meaning. Birds that have been taught to produce words, phrases, or sounds on cue are also best described as mimics. Mimicry crosses over into imitation when birds use learned words in the same way that humans use them. A parrot that can label different types of cup with "cup" is imitating the human use of this word. Some scientists consider that for birds to truly imitate human speech they must produce words in a similar way and be able to recombine the component parts of human speech to produce new words or phrases.

Requiring parrots to articulate words in the same way as humans appears to be setting the bar too high given anatomical differences. Human speech is a two-part process, with sound produced in the larynx subsequently modified as it passes through the pharyngeal, oral, and nasal cavities of the vocal tract. Movement of mobile organs such as the tongue and lips changes the configuration and hence acoustic properties of the vocal tract. Sound vibrations are selectively enhanced or attenuated, producing relatively high sound levels at certain frequencies. These peaks are called formants. Human vowels have distinctive formants, and these allow us to readily distinguish between vowel sounds. Parrots produce sound in the syrinx, an organ somewhat analogous to the human larynx, but situated at the base rather than the head of the trachea. The complex vocalizations produced by parrots were thought to originate in the syrinx, with little modification of sound occurring in the vocal tract. Accordingly, talking parrots were thought to achieve similar acoustic effects to humans using different mechanisms.

Recent research suggests the mechanics of word production in parrots is more similar to humans than previously thought. The parrot vocal tract is now thought capable of modifying sound, with tongue movements likely to be especially important. Gabriël Beckers and colleagues found that by artificially manipulating the tongues of Monk Parakeets *(Myiopsitta monachus)* they could produce complex formant patterns. Such lingual articulation may contribute to the talking ability of parrots, with parrots capable of producing at least some of the formants associated with human vowel production. The relative role of the vocal tract and syrinx in shaping parrot vocalizations remains to be explored. Gabriël Beckers also identified spectral patterns suggestive of formants in the contact calls of Monk Parrots, and raised the possibility that lingual articulation may support a speech-like formant system in natural parrot vocalizations.

A mechanical ability to mold sound in similar ways to humans only partly explains the capacity parrots have for reproducing elements of human speech. The vocal tracts of humans and parrots are more dissimilar than the vocal tracts of humans and other primates. Why can't chimpanzees talk in the same way parrots do? Despite its name, the primary functions of the vocal tract are eating and breathing. Its use for complex communication requires neural structures that allow fine-level control of vocal organs and coordination of breathing and sound production. These evolved separately in parrots and humans, but are apparently lacking in nonhuman primates.

The complexity of human language is said to set it apart from the vocalizations of other animals. Human words are composed of subunits called phonemes and morphemes. Phonemes are sounds within a language that are distinguishable by users and give meaning to words. Thus the phonemes /b/ and /m/ allow us to distinguish between "bat" and "mat." Morphemes are the smallest meaningful or grammatical component of words. The morphemes of "walked" are "walk" and "-ed." Humans have an ability to recombine these subunits to form novel words and to combine words to produce novel sentences. Animals taught a human communication code were thought incapable of understanding the way in which words and sentences were constructed, preventing novel utterances and thus true imitative behavior.

Irene Pepperberg's work with Alex the Grey Parrot *(Psittacus erithacus)* raises the possibility that talking parrots possess some phonological awareness. Alex could recognize the difference between "tea" and "pea," making specific requests for each of these objects. He would practice words, reeling off strings of "mail, chail, benail" when being taught "nail." Like humans, Alex pronounced the phoneme /k/ slightly differently in "corn" and "key." These examples suggest he understood that words were made of subunits. He produced a novel vocalization from subunits, combining "s" and "wool" to produce an approximation of "spool." He later spontaneously perfected this label by adding a /p/ and adjusting the sound of the vowel. Alex's ability developed after nearly 30 years exposure to English speech and specific phoneme training. While Alex used phrases such as "want X" to request various things, he could not combine the large number of labels he had learnt into more complex phrases or sentences with novel meanings. Irene Pepperberg cautions that Alex's utterances should not be considered language, and it is unlikely that humans and parrots will be sitting down discussing the weather any day soon.

Alex's ability to request objects meant that he "failed" the string-pulling task used to test for insight in animals. Irene Pepperberg presented this task to Alex and three other Grey Parrots from her laboratory. Alex and Griffin, another parrot capable of requesting objects, made no physical attempt to retrieve treats suspended below their perch. Instead, they requested the object from their trainer, eventually ignoring it altogether. The other two birds had no trouble completing the task in the expected manner. Alex and Griffin did not fail the test. In fact, the reverse was true. They showed great intelligence by using communication to solve the problem. It is not clear why Alex and Griffin did not try to retrieve the object themselves when their request for assistance was denied. Irene Pepperberg suggests intensive vocal training may have developed parts of the brain dealing with communication at the expense of those dealing with physical coordination.

We should not be surprised by the communicative competence exhibited by Alex given the behavior of wild parrots. Mated pairs of some species are known to sing duets that have a specific phonology and syntax. Work by Christine Dahlin, a graduate student with Tim Wright, has shown that duets in Yellow-naped Amazons *(Amazona auropalliata)* comprise contact calls and three other sex-specific note types. Distinct patterns exist in the sequence of notes within a duet, including the alternate delivery of sex-specific notes. Duets vary in the number of notes they contain, the extent to which note types are repeated or omitted, and the acoustic properties of notes. Many parrots, including Budgerigars *(Melopsittacus undulatus)* and Yellow-naped Amazons, produce rambling vocalizations that comprise a variety of note types and may include imitations of local environmental sounds. These songs are given in a range of contexts, which makes interpreting their function difficult. However, their complex nature means they have potential to transmit environmental information.

The evolution of vocal learning in parrots has an interesting side effect. A Sulphur-crested Cockatoo *(Cacatua galerita)* called Snowball has become famous via a series of YouTube videos showing him dancing to rock or pop songs. These videos came to the attention of researchers, particularly Ani Patel, who noted that Snowball's dancing was well synchronized to the musical beat. They followed up on this observation by manipulating the tempo of one of Snowball's favorite tracks, the Backstreet Boys' "Everybody," and analyzing his response. They found that altering the song's tempo resulted in Snowball slowing or increasing his head bobbing to maintain synchronization. Moving in synchrony with music had been thought a uniquely human trait, but Snowball demonstrated that at least some nonhuman animals have this ability. Trawling through YouTube, another group of re-

Snowball. Irena Schulz, Bird Lovers Only Rescue Service, Inc.

searchers led by Adena Schachner identified 14 parrot species and one elephant species that had similar abilities. Synchronization to a musical beat appears to be restricted to species with a capacity for vocal mimicry, but is not universal across this group. Social complexity may be a factor in this regard. Synchronization of movement with sound has not been observed in wild parrots.

Do parrots play?

The Kea *(Nestor notabilis)* is often described as the clown of the parrot world and has a reputation for mischievous and destructive behavior. The opportunity to investigate how such behaviors evolved attracted University of Nebraska scientists Judy Diamond and Alan Bond to New Zealand. This husband and wife team spent four years studying the foraging ecology and social behavior of Kea, with most of their research undertaken at a rubbish dump within Arthur's Pass National Park. To allow the recognition of individual Kea, birds were banded with numbered metal bands and colored plastic bands. However, catching birds for banding was not easy. The Kea used their considerable intelligence to devise ways of stealing bait from traps without getting caught. The scientists struggled to stay one step ahead of the wily birds, but ultimately 52 individuals were caught and banded over the course of the study. The findings of this research were explored in their book *Kea, Bird of Paradox: The Evolution and Behavior of a New Zealand Parrot* (1999).

Kea were found to play throughout the day, with bouts lasting from a few minutes to an hour. Social play was categorized as either tussle or toss play. Tussle play involved several behaviors. Birds would face each other and jump into the air while flapping their wings, or a bird would roll onto its back while another leapt onto its stomach. Wrestling was often incorporated into tussle play, birds locking bills and pushing at each other with their feet. Toss play involved throwing stones or sticks vertically into the air, and incorporated elements of tussle play such as jumping into the air and wrestling. Play behavior varied with sex and age group. Tussle play was common among young birds and between older females and younger males. Toss play was largely restricted to older birds, most commonly by females soliciting attention from males.

Apart from social play, young Kea spent considerable time playing with objects. Dumped rubbish ensured a variety of potential playthings, but objects that could be probed or manipulated in a variety of ways were preferred. The destruction of objects was a favorite activity, and a group of young birds dismantled a discarded armchair over several days. The attractiveness of objects varied depending on environmental context. A rock that could be rolled down a slope had more potential as a plaything than a similar rock on level ground. The curiosity and playfulness of young Kea have the potential to get them into trouble. Humans have introduced a range of potentially lethal substances into the environment, and young Kea are just as likely to explore fiberglass insulation bats as they are kitchen waste. Conflict arises when birds turn their attention to motor vehicles or expensive camping equipment.

Other parrots do not appear to play as much as Kea, though perhaps people are simply not around when play occurs. Play behavior can be energetically costly, so individuals are thought to benefit from play in some way. Judy Diamond and Alan Bond suggest that social and ecological factors are responsible for differences in the intensity and complexity of play between species. Delayed maturity and the ability of juveniles to source food from adults mean that young Kea have time to play. Social play provides a safe mechanism for young birds to learn their place in Kea society and facilitates cohesion in multiaged flocks. Playing with objects is an extension of the innate curiosity of Kea, a trait that evolved to enable the rapid identification and exploitation of novel food items. Play may also enhance the cognitive development of young birds and thus improve their long-term ability to cope with an unpredictable environment. Toss play appears to be associated with courtship or maintenance of the pair bond and is reminiscent of behaviors used by birds to expose food resources.

Comparisons between species that differ in their ecology, life history, and social behavior have helped unravel the significance of play in parrots. In conjunction with colleagues from New Zealand, Judy Diamond and Alan Bond studied wild Kaka *(N. meridionalis)* and captive Kakapo *(Strigops habroptila)*. Social play in Kaka was limited to young birds, with bouts shorter and less complex than those of Kea. Kaka were not observed to play with objects. Differences between Kaka and Kea were attributed to the former species occupying more benign habitats, living in simpler social groups, and spending less time as juveniles in the company of adults. Nevertheless, there were many similarities in the behavior of these closely related species. Comparisons with Kakapo provided a starker contrast. Kakapo were exceedingly gentle in their play, nibbling or nudging each other. Birds regularly failed to respond to play invitations and bouts were of short duration. Object play was not observed. Social organization in Kakapo is not well understood, though birds are thought to lead a mostly solitary existence. Certainly, they do not form large, multiaged groups like Kea and Kaka. The simple play exhibited by Kakapo provides additional evidence for the importance of sociality in the evolution of complex play behaviors.

Do parrots fight?

Parrots will fight to protect their nest site. Katherine Renton observed interactions between macaws in the vicinity of three Blue-and-yellow Macaw *(Ara ararauna)* nests situated on the edge of a *Mauritia* palm swamp in the Manu Biosphere Reserve, Peru. Most interactions were between breeding and nonbreeding pairs of Blue-and-yellow Macaws. These were typically of low-intensity, breeding birds displaying their yellow underwing toward intruders or displacing them from perches. High-intensity interactions involving aerial chases or physical contact were rare, occurring only when intruding birds perched at the entrance to the nest hollow. Breeding pairs were successful in excluding nonbreeding birds from the nest area, establishing territories that extended some 80 meters from the nest tree. The situation changed dramatically when a predator killed one member of a breeding pair. High-intensity interactions escalated, with an intruding nonbreeding pair eventually taking possession of the nest hollow that was now held by a single bird.

Where there is intense competition for resources, fighting between individuals is more likely. Iron Range National Park on Australia's Cape York Peninsula is home to the charismatic Palm Cockatoo *(Probosciger aterrimus)* and Eclectus Parrot *(Eclectus roratus)*, which nest in woodland and rainforest habitats respectively. It also supports large numbers of Sulphur-crested Cockatoos *(Cacatua galerita)*, whose breeding habitat overlaps that of the other parrots. All three species use large tree hollows for nesting, the scarcity of which results in stiff competition within and between species. Palm Cockatoo males regularly engage in vigorous fights, birds coming together on the wing and occasionally falling to the ground. Combatants suffer significant feather loss in these encounters, but there are no reports of birds dying. Fighting among female Eclectus Parrots is more deadly. Researchers found the body of a female in her hollow exhibiting extensive head wounds following a lengthy fight with another female. The corpses of ten other females with wounds consistent with fighting were found in the vicinity of nest trees. Female Eclectus Parrots also fight with Sulphur-crested Cockatoos over nest hollows, birds attacking and chasing each other. Sulphur-crested Cockatoos often win these skirmishes, preventing or delaying breeding by Eclectus Parrots.

Conflict at nest sites can result in the loss of eggs and the death of nestlings. Galahs *(Eolophus roseicapillus)* routinely interfere in the breeding activities of other cockatoos. This small pink-and-grey cockatoo maintains a presence at nest sites throughout the year and does not tolerate other hollow nesting birds. Its hollow preferences overlap with those of larger black-cockatoos, and disputes over ownership can result in the loss of black-cockatoo eggs. On the western slopes of New South Wales, Galahs may disrupt Glossy Cockatoo *(Calyptorhynchus lathami)* breeding attempts despite each species favoring different hollow types and there being little overlap in breeding seasons. This occurs when the Glossy Cockatoo nest is situated in close proximity to an actual or potential Galah nest hollow. It's not always possible to tell whether the loss of eggs and nestlings is accidental or deliberate on the part of intruders. Scientists sometimes leave artificial eggs in nests that have a history of egg loss in an effort to gain evidence (in the form of a bill imprint) that will help establish the culprit's identity and intentions. Autopsies of dead nestlings may reveal injuries that can be attributed to a known intruder (e.g., puncture wounds that match an intruder's bill shape). Video monitoring of nests has the greatest potential to shed light on these types of events, but it remains underutilized in studies of parrot reproduction.

In some cases, disruption of breeding within or between species appears unrelated to immediate competition for nest hollows or the defense of nest sites. Studying the breeding biology of Crimson Rosellas *(Platycercus elegans)* in Canberra, Elizabeth Krebs found that more than a quarter of all clutches (sets of eggs) were destroyed by other rosellas. There was an excess of nest boxes in the study area, indicating that competition for this

Female Eclectus Parrots *(Eclectus roratus)* will often sit at the entrance to their hollow, calling to let other females know it's occupied. Heinz Lambert

resource was unlikely to be a factor. A study of captive Eastern Rosellas *(P. eximius)* found that dominant females were responsible for the destruction of subordinates' clutches. Dominant birds will benefit from such behavior if it serves to space out nests and reduce competition for food resources, or where it reduces the attractiveness of the nest area to predators due to lower prey densities. Such behaviors evolve under natural conditions over the long term, making the interpretation of results from short-term studies in modified landscapes difficult.

How do parrots avoid predators?

Regent Parrots *(Polytelis anthopeplus)* along the middle Murray River have learnt that flying at speed and sticking to cover reduces the likelihood of attack from Peregrine Falcons *(Falco peregrinus)*. They make daily movements between riparian Red Gum *(Eucalyptus camaldulensis)* forests and adjoining mallee woodlands. These habitats are separated by cleared agricultural land that retains vestiges of the original vegetation along roadsides and in the corners of paddocks. Parrots cross this open country in small flocks, moving between patches of vegetation and following linear remnants wherever possible. They fly within or alongside woodland vegetation at high speed. Open areas are crossed at ground level in contour-hugging flights. If roadside vegetation disappears on the west side of the road but commences on the eastern side, parrots will cross the road at this point and continue their journey. Even raptors adept at hunting in wooded environments, such as Brown Goshawks *(Accipiter fasciatus)*, may have trouble attacking a tight flock of Regent Parrots racing through the canopy.

Predation risk may differ with habitat type and this can be reflected in the differential use of habitats by parrots. Glossy Cockatoos *(Calyptorhynchus lathami)* in central New South Wales share their forest home with Brown Goshawks, Little Eagles *(Hieraaetus morphnoides)*, and Wedge-tailed Eagles *(Aquila audax)*. I observed an unsuccessful attack by a Little Eagle on a foraging group of Glossy Cockatoos. The eagle appeared to locate the group by soaring and then launched a surprise attack below the canopy. This attack might have been successful if the cockatoos had not taken off and flown directly over my head, causing the eagle to veer away when it became aware of me. I found that Glossy Cockatoos spent less time feeding in sites with limited canopy cover, an apparent attempt to limit predation risk. Sites with abundant food that lacked trees were avoided. Conversely, ground-feeding parrots will often avoid areas of dense cover because it limits their capacity for the early detection of predators.

Galahs are more vigilant when foraging in long grass or where their vision is obstructed by woody vegetation. The vigilance behavior of Crimson Rosellas *(Platycercus elegans)* does not appear to alter between closed and open habitats, though birds in open habitats spend a greater proportion of their time feeding. This suggests they feel safer in open environments, though Jennifer Boyer and her coauthors raise the possibility of birds foraging more intensively in these areas so they can leave as soon as possible. Birds respond in several ways when predators are detected. They may carry on with what they were doing, but increase their alert level. They may interrupt their foraging and fly to the nearest cover. Or they may mob the predator. Mobbing is infrequently reported in parrots. In the Peruvian Amazon, Charlie Munn observed macaws mobbing an eagle, the birds circling above the predator and calling loudly. More commonly, parrots disturbed by raptors will take flight and circle skyward in a tight flock without actually approaching the predator.

Cockatoos are said to post sentinels to warn foraging flocks of danger. Individual birds may fulfill this role, but are unlikely to have been allocated the task. Nevertheless, the concept has wide appeal and the term "cockatoo" is Australian slang for a lookout posted at illegal gaming establishments. As is the case with many animals, parrots utilize other species to warn of approaching danger. Foraging Hooded Parrots *(Psephotus dissimilis)* commonly associate with Black-faced Woodswallows *(Artamus cinereus)*. Researchers attempting to collect Hooded Parrot ground-feeding

Budgerigar *(Melopsittacus undulatus)* flock, western Queensland. Lindsay Cupper

Male Red-winged Parrot *(Aprosmictus erythropterus)* feeding on wattle seed, Bowra Station, Queensland.
Dean Ingwersen

Golden-shouldered Parrot *(Psephotus chrysopterygius)*. Patrick Kelly

Pink Cockatoo *(Lophochroa leadbeateri)* nestlings, northwest Victoria.
Victor G. Hurley

Adult male Gang-gang Cockatoo *(Callocephalon fimbriatum)*, southern New South Wales. Matt Cameron

Rainbow Lorikeet *(Trichoglossus haematodus)*, southeast Queensland. Matt Cameron

Male Superb Parrot *(Polytelis swainsonii)*, Wagga Wagga, New South Wales. Matt Cameron

Orange-bellied Parrot *(Neophema chrysogaster)* feeding on glasswort. Chris Tzaros

Swift Parrot *(Lathamus discolor)*. Chris Tzaros

Female Eclectus Parrot *(Eclectus roratus)* captured at Iron Range National Park, Cape York. Matt Cameron

Crimson Rosella *(Platycercus elegans)* color forms—Crimson, Yellow, and Adelaide (top to bottom) (Australian National Wildlife Collection). Matt Cameron

Budgerigars *(Melopsittacus undulatus)* showing "glowing" plumage under UV illumination (mounted specimens). Justin Marshall, University of Queensland

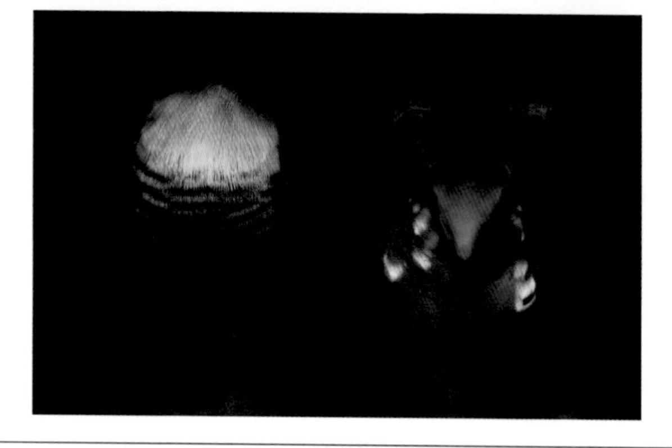

Budgerigar head under white light (top) and UV illumination (bottom). Justin Marshall, University of Queensland

Bald Parrot ***(Pyrilia aurantiocephala)*****.**
Heinz Lambert

Vasa Parrot ***(Coracopsis vasa)*****.**
Heinz Lambert

Burrowing Parrots ***(Cyanoliseus patagonus)*****.** Mauricio Failla

Mealy Amazon *(Amazona farinosa),* Tambopata River, Peru. Green plumage may serve as camouflage.
Matt Cameron

Night Parrot *(Pezoporus occidentalis)* carcass found by the side of the road in western Queensland, 1990.
Australian Museum

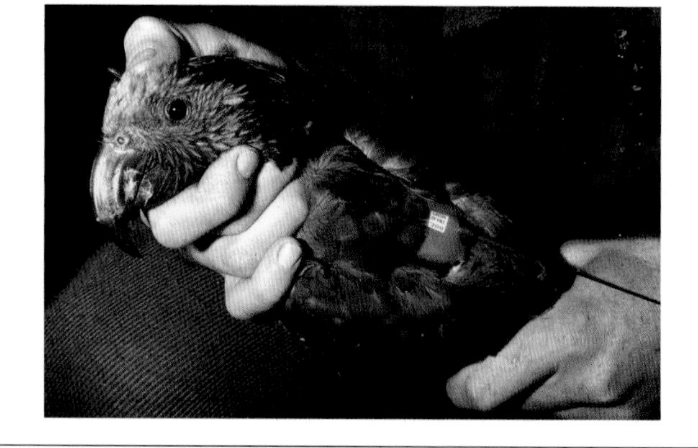

Adult male North Island Kaka *(Nestor meridionalis)* with radio transmitter attached. Terry Greene, New Zealand Department of Conservation

observations were frustrated by the birds flushing whenever the woodswallows gave alarm calls. If the woodswallows moved the parrots tended to follow, suggesting that Hooded Parrots used Black-faced Woodswallows as sentinels.

Predation at the nest is a major source of mortality in many bird species, and cavity nesting is thought to have evolved as a mechanism for reducing such losses. Nevertheless, a range of mammals, birds, and reptiles feed on parrot eggs and nestlings and occasionally catch adults at the nest. Parrots may reduce predation risk by modifying their behavior. Adults are often wary around the nest, anxious not to attract the attention of predators. In particular, smaller species will sit silently and make sure the coast is clear before flying to the nest hollow. Some tropical parrots stop spending the night in the hollow as soon as nestlings are capable of regulating their own temperature. Such behavior has been interpreted as an attempt by females to reduce the likelihood of predators trapping them in the hollow. Glossy Cockatoo nestlings refrain from begging until the female lands at the hollow entrance. This may prevent predators that hunt by sound (e.g., Spotted-tailed Quoll *Dasyurus maculatus*) from pinpointing the nest location. I have often been amazed by the ability of a large, hungry Glossy Cockatoo nestling to remain silent while its mother, who had announced her arrival some time ago, relaxes in a nearby tree, apparently in no hurry to fly to the hollow and feed her offspring.

Snakes are an important nest predator in the tropics and a significant cause of nest failure in some parrots. It has been suggested that parrots respond to snake predation by selecting nest hollows that are inaccessible to snakes or by nesting in areas where snakes are absent or are scarce. That parrots regularly nest in large old trees that lack bark or stand clear of the canopy is often cited in support of this hypothesis. However, this pattern may simply reflect the greater propensity of these trees to form suitable hollows. I found that Glossy Cockatoos in central New South Wales were more likely to nest in areas having a relatively high density of hollows. This may reduce predation risk as predators would need to search a large number of hollows to locate active nests. It is also possible birds are attracted to such areas because of the greater likelihood of finding a suitable hollow in proximity to other breeding birds, Glossy Cockatoos preferring to nest in loose aggregations.

Evidence that parrots are deliberately choosing nest sites to avoid predators can be difficult to collect given the many factors that influence nest site selection. One line of evidence comes from pairs attempting to move from a nest that is prone to predation to one that is relatively safe from predators. Timothy Bonebrake and Steven Beissinger found that Green-rumped Parrotlets *(Forpus passerinus)* do just that. Green-rumped Parrotlet

Hyacinth Macaw *(Anodorhynchus hyacinthinus)* at entrance to nest hollow. Heinz Lambert

nests situated over deep water or clear of vegetation appear to be less vulnerable to predators and fledge more young as a result. These high-quality nest sites are much sought after, with pairs competing among themselves for ownership. Importantly, pairs that change nest site tend to move from low-quality sites to high-quality sites, not vice versa.

Parrots often roost on slender branches in the dense outer foliage of the canopy, where they are less vulnerable to predators. Hanging-parrots roost upside down, ready to make a quick getaway if disturbed by nocturnal predators. Some species roost in termite mounds or tree hollows, benefiting from the equable microclimate and increased protection afforded by such structures. The extinct Carolina Parakeet *(Conuropsis carolinensis)* roosted communally in tree hollows and is known to have roosted in human structures, including barns and log cabins. Roosting parakeets apparently hung by their bill, with some observers suggesting the feet provided additional support. In his monograph on the species, Noel Snyder notes that roosting birds would have been vulnerable to predators such as Raccoons *(Procyon lotor)*. He raises the possibility that Carolina Parakeets may have been unpalatable to predators, possibly due to chemical defenses derived from preferred food such as cocklebursc.

The structure and location of communal roosts suggest they are chosen on the basis of predator avoidance. On Caribbean islands, Brown-throated Parakeet *(Aratinga pertinax)* roosts are located in dense or inaccessible vegetation such as cactus thickets and mangroves. Many parrot roosts are situated along forest edges, be they natural or man made. In southwest Western Australia, Red-tailed Cockatoos *(Calyptorhynchus banksii)* could roost in vast tracts of forest, but choose clumps of remnant vegetation amid farmland or along roadsides. When they roost within forested areas, they select sites adjacent to expansive areas of rock devoid of vegetation. Well-defined edges ensure birds have easy access to the roost, which may help avoid injury in the event birds are disturbed and need to exit and then re-enter the roost in the dark. Orange-fronted Parakeets *(Aratinga canicularis)* shift roosts nightly, while Orange-chinned Parakeets *(Brotogeris jugularis)* move roosts every few weeks. Regular changes in roost location may allow parrots to stay one step ahead of potential predators.

Chapter 5

Parrot Ecology

Do parrots migrate?

The most famous parrot migrants hail from Australia. The Orange-bellied Parrot *(Neophema chrysogaster)* is a small bright-green bird with a blue forehead and orange abdomen. Fewer than 50 individuals remain in the wild, making it one of the world's rarest birds. Orange-bellied Parrots breed in coastal southwest Tasmania, nesting in patches of forest located within or adjacent to buttongrass plains. The breeding period extends from November to mid-February. Once breeding is completed, the entire population migrates north along the Tasmanian coast, eventually crossing Bass Strait to spend the nonbreeding period on mainland Australia. Islands off the northwest tip of Tasmania are important stopover points, and some birds spend the winter on the largest of these, King Island. On mainland Australia, birds can be found close to the coast, favoring sheltered, low-lying habitats such as saltmarshes, coastal wetlands, and pastures. Reaching mainland Australia requires a journey of approximately 600 kilometers, and some birds travel further as they spread out along the South Australian and Victorian coastlines.

When Orange-bellied parrots are breeding in southwest Tasmania, Swift Parrots *(Lathamus discolor)* are similarly engaged in the southeast of the island. Larger and more colorful than the Orange-bellied Parrot, the Swift Parrot is similarly threatened with extinction. Swift Parrots nest in eucalypt forests north and south of Hobart, the locus of breeding activity shifting between seasons as birds look for flowering eucalypts in proximity to nest hollows. Swift Parrots disperse out of the southeast following breeding, seeking feeding opportunities in Tasmania prior to crossing Bass Strait

Orange-bellied Parrot *(Neophema chrysogaster)*. Chris Tzaros

to the mainland. Their winter range extends from southeast Queensland, through coastal and central New South Wales, across Victoria into South Australia. The location of birds within this winter range depends on food availability. Birds located in the far north of the range have traveled in excess of 2,000 kilometers, the longest migration of any parrot.

The dispersal of Swift Parrots from their breeding areas coincides with a decline in flowering of key feed trees such as Tasmanian Blue Gum *(Eucalyptus globulus)*. Crossing Bass Strait gives them access to the box-gum woodlands on the inland slopes of the Great Dividing Range, where eucalypts provide a ready source of nectar and lerp (the sugary shell produced by sap-sucking insects). When drought reduces the productivity of inland areas, food is often available in wetter coastal regions. Orange-bellied Parrots benefit from their migration in a similar way, with occupation of their nonbreeding habitat coinciding with periods of peak seed production in their favorite food plants (e.g., glassworts). Why don't Orange-bellied and Swift Parrots breed on the mainland? The answer lies in the patchy temporal and spatial distribution of food resources. Birds are constantly on the move as they track food availability, and this precludes them settling in one area for an extended period.

The Orange-bellied and Swift Parrots are classic migratory species. They exist as single populations that make seasonal movements between locations. The situation is more complicated for the Blue-winged Parrot *(Neophema chrysostoma)*, another Tasmanian migrant and close relative of the Orange-bellied Parrot. The Blue-winged Parrot occurs as two separate populations that spend the spring-summer breeding period in Tasmania and southern Victoria respectively. In autumn, Tasmanian birds cross Bass Strait to the mainland. At this time, the species' range expands into inland southeastern Australia. Two scenarios have been proposed to explain this shift in distribution. The first involves migrating Tasmanian birds leap-frogging those resident in southern Victoria. The second entails birds in southern Victoria moving inland, with vacated habitats being occupied by Tasmanian birds. At least part of the Tasmanian population is thought to over-winter on Tasmania.

In Guatemala, Robin Bjork studied the movements of Mealy Amazons *(Amazona farinosa)* using a combination of ground-based and aerial radio tracking. She found that in the north of the country, a large proportion of the population engaged in predictable migrations during the nonbreeding season. Once breeding was completed, birds traveled northeast to sites 80 kilometers from their breeding area. They spent three to four weeks at these sites before returning to the breeding area for a few weeks. They then traveled southwest to sites up to 190 kilometers from the breeding area, where they remained for a few months before returning to the breeding area. These movements were thought to be related to the availability of food resources, birds migrating from areas having low fruit abundance to areas having higher fruit abundance. When a favorite food was declining in availability in the breeding area, it was abundant in forests to the north-east. Similarly, when birds were in the southwest they had access to fruiting trees that did not occur within the breeding areas. A number of other parrot species vacated the breeding area at the same time as the Mealy Amazons. The movement patterns of these species were unknown, though

Mealy Amazon *(Amazona farinosa)* with radio tracking collar attached.
Robin Bjork

View from an aircraft being used to track Mealy Amazons *(Amazona farinosa)* in Guatemala. Note the antennae attached to the wing strut, and the Mayan temples on the ground. Robin Bjork

they did not appear to follow the same habitat use pattern displayed by the Mealy Amazons.

In the Amazon, changes in parrot abundance over time suggest birds are moving in and out of areas. On the Tambopata River in southeastern Peru, total parrot abundance shows a distinct annual pattern that is correlated with local fruiting and flowering patterns. Parrot numbers peak in the late dry season through the early wet season. This is a period of maximum food availability, wind-dispersed seeds ripening late in the dry season and fleshy fruit production peaking early in the wet season. A number of biologists have reported on seasonal fluctuations in the abundance of large macaws within their study areas. In lowland forest along the Tambopata River, numbers peak during the wet season, a time when birds are breeding. Similarly, in floodplain rainforest of the Manu Biosphere Reserve, three times as many birds are observed in the wet season compared to the dry season. Contrastingly, in upland forests in Ecuador, macaws are more common in the dry season compared to the wet season. Taken as a whole, these observations suggest that birds breed in lowland forests during the wet season and then move out of these habitats during the dry season. It remains to be seen whether dry season movements are directed toward particular sites, or simply entail birds wandering in search of food.

In Africa, the available information on parrot movements is largely anecdotal. There is little evidence for regular seasonal migration in most populations, though local movements or nomadic wanderings are commonly reported. One exception is the Brown-necked Parrot *(Poicephalus fuscicollis)*. In at least the southern parts of its range, it undergoes post-breeding seasonal movements. Craig Symes studied Greyheaded Parrots *(P. f. suahelicus)*, the southern subspecies of the Brown-necked Parrot. The work was part of an MSc undertaken through the then University of Natal. His supervisor was Mike Perrin, who has made an important contribution to the study of many African parrots. Symes was well qualified to undertake the work, having previously assisted Olaf Wirminghaus with his research on the closely related Cape Parrot *(P. robustus)*. At one site in northeastern South Africa, he found that Greyheaded Parrots turned up regularly in August each year. Their arrival coincided with the setting of fruit in the Mabola Plum *(Parinari curatellfolia)*, a relatively nutritious fruit that dominated the diet during the period following the breeding season. At the same time, food availability at a breeding site 100 kilometers to the west declined. Further research may reveal regular seasonal movements in other African parrots.

The conservation of parrot populations that undertake large-scale movements is challenging. Even large reserves may not accommodate movements of a few hundred kilometers. There is no guarantee that large reserves will encompass all the resources used by a population. Understanding the nesting requirements of parrots is relatively straightforward. As a consequence, parrot conservation efforts often focus on the protection of breeding habitat. However, there is a need to identify and secure other critical resources (e.g., nonbreeding season foraging habitat) if extinction is to be avoided. Corridors or stepping stones that facilitate access to these resources must also be protected. This requires an understanding of movement patterns, information that can only be gathered through detailed investigation.

The pattern of winter habitat use by the Swift Parrot provides a good example of the complexities involved in conserving migratory species. In 2002, Swift Parrots were most abundant on the New South Wales central coast, while in 2004 the center of activity was 800 kilometers to the southwest in central Victoria. Birds are found in tracts of continuous forest one year, but occur in small remnants or scattered trees the following year. Despite 15 years of survey effort, Swift Parrots still have the capacity to surprise, going missing in some years before turning up at unexpected locations. Such spatial and temporal variability in the selection of foraging sites means that identifying and protecting habitat are problematic. High priority conservation sites are those used repeatedly or known to provide refuge during periods of widespread food shortage. These can only be identified by analyzing long-term data sets. We do not understand how Swift Parrots identify patchily distributed resources over their vast winter range, a knowledge gap with the potential to undermine attempts at creating a network of appropriately managed habitat.

Which geographic regions have the most species of parrot?

Australasia is the biogeographical region with the greatest number of parrot species. The large number of island groups in the region is an important contributing factor to this statistic. Parrots are strong fliers, capable of crossing large expanses of water to reach distant islands. They are known from some of the remotest islands in the Pacific Ocean. If a parrot can establish itself on an island, there is a good chance it will diverge from the parent population through processes such as natural and sexual selection. At some point, parent and daughter populations may become sufficiently different that interbreeding is unlikely if they come into contact. Such reproductive isolation signals the evolution of a new species. It's worth noting that divergence is dependent on the population remaining isolated. Continual exchange of individuals between the parent population and new population will reduce the likelihood of them diverging. Species that occasionally cross water gaps are more likely to evolve into new species (speciate) than those that readily fly between islands.

Island archipelagos will often have one or two representatives from a group of closely related species with nonoverlapping ranges. These groups are termed superspecies, with members referred to as allospecies. They are thought to derive from a common ancestor that dispersed through the region, with isolated populations subsequently diverging. Allospecies may undergo further divergence within an archipelago, resulting in the recognition of subspecies. The lorikeets provide some of the best examples of dispersal-driven speciation in parrots. The genus *Lorius* is distributed from the Moluccas to the Solomon Islands, with the six allospecies occupying mostly separate ranges. Populations within *Lorius* allospecies are also diverging from each other, principally in relation to color patterns. For example, there are three recognized subspecies of Chattering Lory *(Lorius garrulous)*, each occupying different islands in the North Moluccas.

The Neotropics is not far behind Australasia in the richness of parrot

Table 5.1. Parrot regions of the world

Region	Land masses	Number of parrot species*
Neotropical	Mexico Central America South America	153
Afrotropical	Africa Madagascar	24
Indomalayan	India continental Southeast Asia	17
Australasian	Wallacea New Guinea Australia New Zealand Southwest Pacific islands	166

* Some species occur in more than one region.

Table 5.2. Australasian parrot subregions

Subregion	Land masses	Number of parrot species*
Southeast Asian Islands	Wallacea New Guinea Northern Melanesia	110
Australian	Australia Tasmania	53
Southwest Pacific	Southern Melanesia Polynesia Norfolk Island New Zealand	27

* Some species occur in more than one subregion.

species. Climatically induced shifts in the distribution of plant communities during the Pliocene (5.3 to 1.8 million years ago) and Pleistocene (1.8 million years ago to 10,000 years ago) played an important part in the evolution of Neotropical lowland parrots. Ancestral populations became fragmented, isolated by areas of unsuitable habitat. Mountain building in the Andes (Andean uplift) may also have played a role, generating water bodies within Amazonia that further fragmented populations. Molecular techniques have assisted the reconstruction of speciation events. Camila Ribas and colleagues explored speciation in *Brotogeris* parakeets, a genus of eight species distributed throughout Neotropical lowlands. They compared DNA sequences to confirm the validity of species, which had originally been separated on the basis of morphology (appearance), and to determine their evolutionary relationships. Two groups were identified, which did not necessarily reflect key morphological differences or current distributions. Species with short, rounded tails were shared between groups, as were species occurring within Amazonia. Using molecular dating, they calculated how long populations had been separated and estimated the dates when this occurred. The two groups diverged between five and six million years ago, with subsequent splits occurring through the Pliocene and Pleistocene.

Andean uplift played a direct role in generating parrot diversity in the Neotropics. Camila Ribas and colleagues investigated the evolutionary and geographical history of *Pionus* parrots, a genus of seven species occurring in either mountain or lowland habitats. The three mountain species were found to be more closely related to lowland species than they were to each other. Splits between mountain and lowland species coincided with periods of mountain building, suggesting that mountain species were transported to higher elevations as part of this process. Unsuitable habitat on the lower slopes served to separate mountain and lowland populations. Through adaptation to mountain habitats, mountain lineages diverged to produce modern subspecies. The timing of diversification within mountain clades is consistent with Pleistocene climate change. Mountain populations contracted to higher elevations during warm interglacials, where they were isolated from each other by valleys or tropical forest. Outside the Neotropics, mountain ranges supporting habitats favored by parrots are rare or limited in extent.

How many parrot species live in an area?

The diversity and abundance of parrots in Neotropical forests are two of their defining characteristics. At the Tambopata Research Center in southeastern Peru, 21 species of parrot have been recorded from a relatively small area of lowland forest. Researchers regularly record 17 species, while the casual visitor will encounter a dozen different parrots without too much effort. Many of the occurring parrot genera are represented by two species, with the macaws having six representatives. The situation is similar in southeast Brazil, with 15 species of parrot recorded from a 45,000-hectare remnant of lowland Atlantic Forest. The ways lowland species partition resources are not well understood. Differences in body size may reduce competition for food and nest sites, while closely related species tend to favor different habitats. Mountain forests support fewer parrots. Closely related mountain species typically occur at different elevations, though there is often considerable overlap in altitudinal ranges. For example, Bronze-winged Parrots *(Pionus chalcopterus)* range between 1,400 to 2,400 meters, while Speckle-faced Parrots *(P. tumultuosus)* occur from 2,000 to 3,000 meters. Parrot species richness can be particularly high when the area under consideration includes lowland and upland environments. This helps explain why countries such as Colombia and Peru, despite their relatively small size, each support around 50 species of parrot.

Australasia is the only other region where local species richness may approach that of Neotropical lowland forests. Harry Bell surveyed a 2.5-hectare plot of lowland forest in New Guinea over a two-year period. He recorded 15 species of parrot, with a further two species added by other ornithologists. Most of the recorded species were resident, though a number of mountain species (e.g., Goldie's Lorikeet *Psitteuteles goldiei)* visited when preferred foods were abundant. One mountain species, Red-breasted Pygmy-parrot *(Micropsitta bruijnii)*, was recorded during a cyclonic disturbance. Andrew Mack and Debra Wright reported on bird species richness

The Tambopata Research Center in southeastern Peru is home to 21 species of parrot. The Tambopata Macaw Project, managed by Donald Brightsmith from Texas A&M University, undertakes long-term research on macaws and other parrots in the surrounding rainforest.

at the Crater Mountain Biological Research Station in the Eastern Highlands of Papua New Guinea. The 260-hectare study area encompassed the transition from lowland to mountain forest. They recorded 19 parrot species, with an additional species added by later researchers. Parrots ranged in size from the tiny Buff-faced Pygmy-parrot *(M. pusio)* to the massive Palm Cockatoo *(Probosciger aterrimus).*

A contributing factor to high parrot species richness at sites in the Neotropics and Australasia may be the paucity of squirrels. Squirrels are specialized seed predators and could be expected to compete with parrots for this critical resource. Richard Corlett and Richard Primack highlighted the inverse relationship between parrot and squirrel species richness in tropical rainforests. New Guinean rainforests lack tree squirrels, while Amazon rainforests have only seven species. It is rare to find more than two or three squirrel species coexisting in South American forests. Conversely, regions with few parrots have lots of tree squirrels. African rainforests support 14 species of tree squirrel, while those on the Southeast Asian mainland have 31 species. Nine species of tree squirrel have been reported from a single area in Gabon.

Table 5.3. Parrot species recorded from a 2.5-hectare plot of lowland rainforest in Papua New Guinea (Bell 1982).

Common name	Scientific name
Buff-faced Pygmy-parrot	*Micropsitta pusio*
Red-breasted Pygmy-parrot	*Micropsitta bruijnii*
Palm Cockatoo	*Probosciger aterrimus*
Sulphur-crested Cockatoo	*Cacatua galerita*
Orange-fronted Hanging-parrot	*Loriculus aurantiifrons*
Yellow-streaked Lory	*Chalcopsitta sintillata*
Dusky Lory	*Pseudeos fuscata*
Rainbow Lorikeet	*Trichoglossus haematodus*
Goldie's Lorikeet	*Psitteuteles goldiei*
Black-capped Lory	*Lorius lory*
Red-flanked Lorikeet	*Charmosyna placentis*
Fairy Lorikeet	*Charmosyna pulchella*
Orange-breasted Fig-parrot	*Cyclopsitta gulielmitertii*
Red-cheeked Parrot	*Geoffroyus geoffroyi*
Blue-collared Parrot	*Geoffroyus simplex*
Eclectus Parrot	*Eclectus roratus*
Papuan King-parrot	*Alisterus chloropterus*

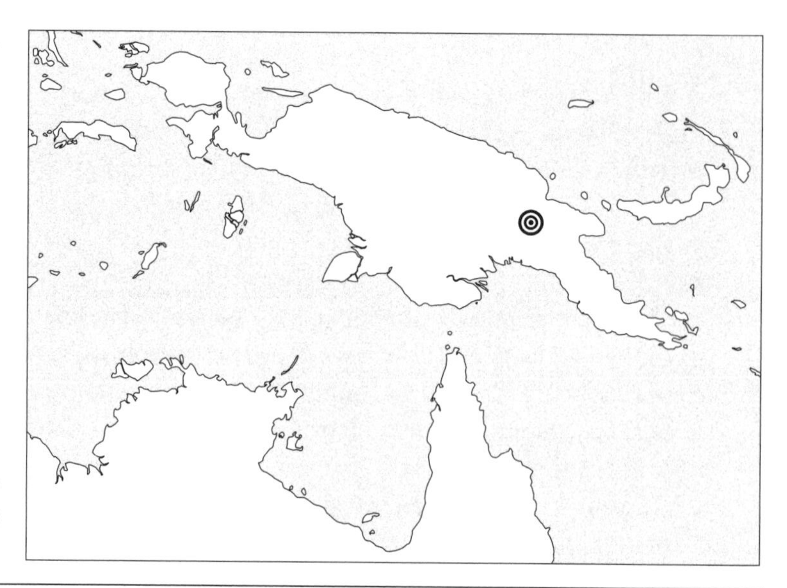

Andrew Mack and Debra Wright recorded 19 species of parrot from 260 hectares of forest at Crater Mountain in Papua New Guinea.

How many parrots live in the Australian desert?

Australia is the land of desert parrots, with eleven species characteristic of the arid zone. The distribution of several other species extends into this area. Providing water is available, parrots are the most abundant birds in the Australian desert. Their dominance is even greater when their large size relative to other birds is taken into account. Parrots occur in deserts on other continents, but don't dominate the bird fauna as they do in Australia. In South America, Monk Parakeets *(Myiopsitta monachus)* and Burrowing Parrots *(Cyanoliseus patagonus)* are found in the lowland desert of western Argentina, while the Mountain Parakeet *(Psilopsiagon aurifrons)* makes its home in the dry valleys of the Peruvian Andes. In Africa, Rüppell's Parrot occupies the arid scrubland of northern Namibia and southwest Angola, a distribution shared with the Rosy-faced Lovebird *(Agapornis roseicollis)*.

Shrublands are important habitat for Australia's desert parrots. Tall acacia-dominated shrublands occupy more than 30% of the arid zone, while low chenopod shrublands are also widespread. Shrublands typically have a grassy understory, and thus provide parrots with a variety of food. Ian Rowley examined the diet of Galahs from acacia shrublands in Western Australia. Acacia, chenopod, and grass seed were all found to be important. This pattern is typical of many desert parrots, which are best described as generalist granivores. Overall, smaller species tend to have less diverse diets than larger species. The Budgerigar *(Melopsittacus undulatus)*, the smallest desert parrot at 30 grams, feeds predominately on grass seed, while Pink Cockatoos *(Lophochroa leadbeateri)*, the largest desert parrot at 370 grams, feed on seeds, fruits, vegetable matter, and invertebrates.

Table 5.4. Australia's desert parrots

Common name	Scientific name
Galah	*Eolophus roseicapillus*
Pink Cockatoo	*Lophochroa leadbeateri*
Cockatiel	*Nymphicus hollandicus*
Australian Ringneck	*Barnardius zonarius*
Bluebonnet	*Northiella haematogaster*
Mulga Parrot	*Psephotus varius*
Bourke's Parrot	*Neopsephotus bourkii*
Scarlet-chested Parrot	*Neophema splendida*
Budgerigar	*Melopsittacus undulatus*
Night Parrot	*Pezoporus occidentalis*
Princess Parrot	*Polytelis alexandrae*

How do parrots survive in the desert?

Parrots have a number of physiological and behavioral adaptations that allow them to conserve water and cope with high temperatures. A Galah's *(Eolophus roseicapillus)* evaporative water loss at low temperatures is less than half that experienced by the Greater Roadrunner *(Geococcyx californianus)*, a common inhabitant of North American deserts. Conversely, Galahs can dramatically increase water loss at high temperatures in order to lower their body temperature via evaporative cooling. This allows them to maintain body temperatures within tolerable limits, even when exposed to high temperatures for several hours. The ability to dissipate heat in excess of that produced through metabolism is shared by other desert parrots such as the Budgerigar *(Melopsittacus undulatus)* and Monk Parakeet *(Myiopsitta monachus)*. The costs, in terms of water loss, associated with evaporative cooling are minimized by the Galah's ability to tolerate higher-than-normal body temperatures. Water loss is further reduced by the capacity Galahs have to produce highly concentrated urine and to reabsorb water and salt from the cloaca, which provides a single opening to the exterior for the digestive, urinary, and reproductive systems. High daytime temperatures and the need to forage away from water sources mean that Galahs are still likely to face water deficits. Galahs are able to tolerate periodic dehydration and can rehydrate themselves from waterholes that are salty if necessary.

Low metabolic rates are one mechanism employed by desert animals to cope with high temperatures. The breakdown, or metabolism, of food produces heat. Low metabolic rates reduce the heat load on animals inhabiting hot environments. Animals with low metabolic rates also need to consume less food, reducing the time spent foraging and thus their exposure to high temperatures. There is limited evidence for low metabolic rates among desert parrots. Christine Cooper compared the metabolic ecology of two Red-tailed Cockatoo subspecies, the Forest Red-tailed Cockatoo *(Calyptorhynchus banksii naso)* and Inland Red-tailed Cockatoo *(C. b. samueli)*. The latter subspecies occupies more arid environments and was found to have a lower basal metabolic rate. A number of other studies comparing desert-dwelling parrots with those from moister habitats have failed to find any relationship between aridity and metabolic rate.

Parrots obtain limited moisture from their seed-based diet and must have access to free-standing water. Small parrots drink in the morning and evening, while larger species visit waterpoints at the end of each day. Surface water is naturally scarce in the arid zone and may be located a considerable distance from foraging areas. Most desert parrots are strong fliers, able to efficiently commute between food and water resources. The mobility of desert parrots allows them to range over large areas, exploiting patchily distributed resources that collectively provide sufficient food to support local populations. The availability of surface water is an important determinant of parrot distribution in the arid zone. Parrots are unlikely to be present where surface water cannot be found in reasonable proximity to food resources. They probably utilize a network of reliable water sources, shifting between them dependent on the spatial distribution of foods. Artificial waterpoints have allowed desert parrots to permanently occupy regions that may once only have been habitable in certain seasons or years, as well as facilitating the spread of more mesic species into the arid zone.

Are there parrots in alpine areas?

The Kea *(Nestor notabilis)* is often said to be the world's only alpine parrot. This is partially correct. While the Kea is at home in the mountains, it can be found at lower elevations. At high elevations it makes use of alpine habitats, but is reliant on adjacent beech forest. Most nests are located within beech forest, and beech trees *(Nothofagus)* are an important source of food, birds feeding on buds, leaves, and nuts. Nevertheless, access to habitats above the treeline is important. Kea dig up Mountain Daisies *(Celmisia)* from alpine grasslands, while bright red Coprosma *(Coprosma)* berries are sourced from alpine shrublands. Kea can often be observed foraging over grassland areas recently exposed by retreating snow patches.

The Kea is not the only New Zealand parrot to exploit the mosaic of habitats associated with the treeline. The last mainland wild population of Kakapo *(Strigops habroptila)* was found near the treeline on the high sides of glacial valleys in Fiordland. Home ranges encompassed beech forest, alpine scrub, and tussock grassland. Snow covered the ground during the winter months, and birds may have sought refuge in beech forests during these periods. Such habitat is today considered less than ideal for Kakapo, with breeding dependent on the availability of podocarps that periodically produce abundant fruit. These typically occur at lower elevations. Historically, some Kakapo populations may have depended on beech seed to raise their young. The remoteness of the Fiordland site probably reduced the impact of feral predators and competitors and allowed the population to hang on longer than other mainland populations.

The Andes run the length of South America, extending from Colombia in the north to Chile in the south. There are extensive areas above the treeline, predominately grassland with patches of shrubland and woodland. The Andes supports a diverse parrot fauna, though most species are restricted to habitats below the treeline. The Speckle-faced Parrot *(Pionus tumultuosus)* occurs at 2,000 to 3,000 meters, but can be found above the treeline in wooded habitats. The Red-billed Parrot *(P. sordidus)* and Bronze-winged Parrot *(P. chalcopterus)* occur at slightly lower elevations. Other *Pionus* are essentially lowland species. The genera *Psilopsiagon* and *Bolborhynchus* are more specialized on mountain habitats, and are often collectively referred to as mountain parakeets. The Mountain Parakeet *(Psilopsiagon aurifrons)*, Andean Parakeet *(Bolborhynchus orbygnesius)*, and Rufous-fronted Parakeet *(B. ferrugineifrons)* are known from areas above the treeline. The Mountain Parakeet has a number of subspecies, one of which *(P. a. margaritae)* occurs in the high Andes between 3,000 to 4,000 meters. The Andean Parakeet and Rufous-fronted Parakeet occur at similar elevations. These small green parakeets are typically associated with patches of woody vegetation or more open habitats supporting scattered shrubs or trees. Mountain Parakeets and Andean Parakeets are thought to descend to lower altitudes during winter.

Mountain habitats in Australia are limited in extent. The principal alpine area is centered on Mt. Kosciusko (2,228 meters) in southeastern Australia. It has continuous snow cover for around four months of the year, and a complex mosaic of alpine heaths, herb fields, and freshwater wetlands dominates the vegetation. A lack of trees and winter snow cover mean alpine areas provide limited habitat for parrots. The subalpine zone is characterized by snow gums, stout eucalypts that rarely exceed 15 meters in height, and is home to a number of parrots. I located around 80 Gang-gang Cockatoos *(Callocephalon fimbriatum)* during a day spent traversing snow gum woodland in autumn, birds revealing their presence by the noise they made crunching snow gum fruit to access the enclosed seed. Crimson Rosellas *(Platycercus elegans)* similarly feed on the fruits of snow gums, while Yellow-tailed Cockatoos *(Calyptorhynchus funereus)* extract insect larvae from the trunks and branches of trees. Canopy foraging allows all three species to maintain a presence in the subalpine zone even when

Adult male Gang-gang Cockatoo *(Callocephalon fimbriatum)*, southern New South Wales. Matt Cameron

the ground is covered with snow. Altitudinal movements have been claimed for all three parrots, with at least part of their populations said to move to lower elevations during winter.

How do parrots survive droughts?

Parrots have two options for dealing with drought. They can move out of affected regions, or they can endure the hard times and await the return of more favorable conditions. Pink Cockatoos *(Lophochroa leadbeateri)* adopt the latter strategy. Found in some of the harshest environments in Australia, they survive by ranging over large areas and feeding on a wide variety of foods. Ian Rowley and Graeme Chapman studied Pink Cockatoo diets on the edge of the Western Australian wheatbelt. They found them to be omnivorous, consuming the seeds and soft parts of herbaceous plants, the seeds of native shrubs and trees, and invertebrates. Low population densities reduce the pressure on food resources, while the wide dispersion of nest sites serves to uniformly allocate food between family groups. Pink Cockatoos' knowledge of water resources would have paralleled that of indigenous people, who were renowned for their ability to cross desert country by navigating between widely spaced water holes. The spatial memories of older flock members no doubt help birds find food and water when conditions are at their worst.

Pink Cockatoo *(Lophochroa leadbeateri)* feeding on wattle seed, Bowra Station, Queensland. Dean Ingwersen

Many of Australia's desert parrots are considered nomadic, wandering in search of food and water and settling to breed where these resources are located in abundance close to breeding habitat. Nomadism and opportunistic breeding are considered adaptations to the unpredictability of the Australian arid zone, though the evidence for such behavior is largely anecdotal. Probably many species wander locally, taking advantage of patchily distributed food resources. These will often be associated with run-on areas, where light rains provide sufficient moisture to cause a flush of growth and allow seed production. In most years, enough food will be available to support parrot populations and allow some breeding. Widespread rain or flooding may result in sustained breeding across a region, local numbers bolstered by birds from other areas. If the food supply gives out, birds must move or die. These movements are directed toward the structurally similar but moister habitats of the surrounding semiarid zone. Food is more likely to be available in the north during winter and in the south during summer, and this imposes a seasonal pattern to movements.

The ecology of inland Australia is one of cyclical boom and bust. Plants and animals hunker down during the dry times and then rush to reproduce when conditions are favorable. This leads to an increase in animal numbers before the inevitable crash once dry conditions return. Parrots do not readily fit this model. Their capacity for population increase is limited by the time taken to rear young to independence, and their mobility allows them to depart regions when conditions deteriorate. Budgerigar *(Melopsittacus undulatus)* numbers can fluctuate widely between years, more so than those of other desert parrots. Large numbers of birds are observed in favorable seasons, and populations are assumed to decline during periods of drought. However, the mechanisms of population growth and regulation have not been studied. Budgerigars feed on a variety of plant seeds, shifting between species depending on abundance, and readily take seed from the ground. They do not require green seed for breeding. Accordingly, they are well placed to take advantage of heavy rains that result in staggered seed production and the storage of large quantities of seed on the soil. Under these conditions, numbers can build up rapidly within a region and widespread breeding occurs.

Charles Barrett witnessed more than one million Budgerigars visiting a waterhole in the Northern Territory over a two-hour period. Up to 20,000 birds were present at any one time. This observation was made in the first half of the last century, though the exact date is unknown, and he did not report on prevailing conditions. Budgerigar numbers may have increased following a run of good seasons, with subsequent drought concentrating these birds on limited water resources. In 2009, huge flocks of Budgerigars were observed in western Queensland near the town of Boulia. Numbers of Budgerigars were linked to heavy summer rains and flooding of inland river systems. Reports were made of 17 pairs nesting in a single tree, and of birds incubating clutches laid on the ground. The relative roles of immigration and breeding in the formation of superflocks remain elusive. We do not know the factors governing the breakup of superflocks, though declining food resources or the emergence of new food resources elsewhere presumably play a role, nor do we have any knowledge of mortality rates of dispersing birds.

Droughts are not restricted to desert regions. Areas having wetter climates may be subject to dry spells. Parrots in temperate regions have the same choices as desert parrots when conditions deteriorate. They can stay put or seek better conditions elsewhere, though the widespread nature of many droughts makes locating unaffected areas difficult. Glossy Cockatoos *(Calyptorhynchus lathami)* in central New South Wales sat tight during the 2002 drought, one of the worst on record due to above average temperatures. They were buffered from the worst effects of the drought by the habit their food plants have of storing seed in the canopy. This old seed was the equivalent of survival rations. It kept birds alive, but was not of sufficient quality to support breeding. Glossy Cockatoos are long-lived and well able to withstand limited or no breeding during drought years, the population recovering when conditions are more favorable for seed production in their food plants. This pattern of population size fluctuating in accordance with cyclical drought is probably typical of many parrot species in southeastern Australia.

Budgerigar *(Melopsittacus undulatus)* flock, western Queensland. Lindsay Cupper

Do parrots get sick?

One of the principal diseases of parrots is Psittacine Beak and Feather Disease (PBFD), also known as Psittacine Circovirus Disease (PCD). The disease is caused by the Beak and Feather Disease Virus (BFDV) and takes its name from beak deformities and feather abnormalities displayed by chronically infected birds. The most common form of transmission is from adults to nestlings, though transmission between free-flying birds is possible. The acute form of the disease is associated with nestlings and causes death in one to two weeks. The chronic form of the disease occurs in older birds and is evidenced by the appearance of abnormal feathers during moult. Birds with chronic PBFD die within two years of the onset of symptoms, usually through contracting secondary infections. All parrots are susceptible, but species differ in their resistance to infection. Cockatoos are particularly vulnerable, though Cockatiels *(Nymphicus hollandicus)* are rarely infected. PBFD is endemic in Australia's wild parrot populations, and has a worldwide distribution in captive parrots. It has been reported in wild Red-fronted Parakeets *(Cyanoramphus novaezelandiae)* in New Zealand, wild Cape Parrots *(Poicephalus robustus)* in South Africa, and in the managed Mauritius Parakeet *(Psittacula eques)* population on Mauritius.

The lack of effective treatment means PBFD is of concern to parrot owners. There is the potential for endangered species breeding programs to be disrupted by the establishment of the virus in captive populations. While the disease is unlikely to pose a threat to healthy wild populations having previous exposure to the virus, it is a threat to populations of conservation concern that are small or have had no prior exposure. The disease is present in the endangered Swift Parrot *(Lathamus discolo*r) population, and deaths of nestling Swift Parrots from the acute form of the disease have been reported. The risk posed to endangered parrots by PBFD has led to its declaration as a key threatening process in Australia and the preparation of a threat abatement plan. Management of the disease is hampered by the lack of a commercial vaccine, the resistance of the virus to most disinfectants, and the ability of asymptomatic birds to spread the disease. Testing is important to identify birds that are infected or may carry the virus. Parrots should not be released into the wild or transferred between facilities unless their BFDV status is known and appropriate quarantine procedures are followed.

Proventricular Dilation Disease (PDD) is a disease known from captive parrots, but which may also exist in wild populations. In the 1970s, it appeared among wild caught macaws imported to the United States and Europe. It is sometimes referred to as Macaw Wasting Disease. It causes dilation of the proventriculus, an organ of the upper digestive tract responsible for secreting digestive enzymes and transferring food from the crop to the gizzard. Impacts on the digestive system cause birds to lose condition, while those on the central nervous system cause problems with balance and walking. All parrots are thought to be susceptible, though it is more common in particular groups or species. PDD is highly infectious and can quickly spread through captive populations. The disease is reported as causing the death of five Spix's Macaws *(Cyanopsitta spixii)* from a captive breeding facility, a significant loss given the critically endangered status of this species.

The cause of PDD was unknown until researchers from the University of California, San Francisco, linked the disease to bornaviruses. The team used a ViroChip to screen samples from birds known to have died from PDD. The ViroChip is capable of identifying all known viruses as well as related unknown viruses. The chip had previously played a role in identifying the cause of Severe Acute Respiratory Syndrome (SARS) in humans. They found a new type of bornavirus was present in the majority of infected birds, but absent from control samples. Subsequent work confirmed the ability of these Avian Bornoviruses (ABVs) to cause PDD in parrots.

To date, management of PDD has proven difficult because of difficulties diagnosing the disease and the potential for asymptomatic birds to act as carriers. Identification of the causative agent of PDD has made possible the development of commercial diagnostic tests. These rely on the detection of bornaviral genetic material or antibornaviral antibodies to diagnose bornavirus infection. Testing will assist in eradicating the disease from captive flocks through the identification and quarantine of affected individuals.

While captive parrots are subject to a range of internal parasites, wild parrots appear remarkably resistant to infestation. Juan F. Masello and coauthors looked for blood and intestinal parasites in a colony of Burrowing Parrots *(Cyanoliseus patagonus).* Despite extensive sampling of adult and nestling birds over a number of breeding seasons, they failed to detect their presence. Similarly, a four-year study of wild Hyacinth Macaw *(Anodorhynchus hyacinthinus)* nestlings in the Brazilian Pantanal could find no evidence of blood or internal parasites. Reviewing the literature on parasite investigations of wild parrots, Juan F. Masello and coauthors could not locate a study where blood parasites had been detected. Intestinal parasites were reported from a range of species, though infestation rates were low at around 30% of cases. They concluded it made sense for parrots to invest heavily in the suppression of parasites given their longevity. Burrowing Parrots would especially benefit from an enhanced immune system due to the increased potential for parasite transmission within nesting colonies.

Parasites are routinely detected in wild parrots brought into wildlife clinics. This may be due to the potential for birds in urban or agricultural environments to come into contact with domesticated animals. During the winter months, juvenile Australian King-parrots *(Alisterus scapularis)* suffering from emaciation, diarrhea, and weakness are routinely referred to wildlife clinics. This syndrome is associated with Spironucleus protozoa, the taxonomy and host specificity of which is unknown. Birds may be especially vulnerable during winter, a time of natural food shortage. Reliance on bird feeders during winter may bring birds into contact with pets or facilitate the spread of the disease. In the Western Australian wheatbelt, Ian Rowley reported a wasting syndrome that afflicted nestling Galahs *(Eolophus roseicapillus).* Birds lost weight, failed to develop normally, and suffered diarrhea. The disease was widespread within the study area, though some areas were more affected than others. In some years, the disease resulted in the death of around 80% of nestlings. Cereal grains dominated the diet of birds within the region, birds often foraging in the company of livestock and feeding on grain that passed through livestock undigested. The cause of the disease was not identified.

Are any parrot species nocturnal?

The Night Parrot *(Pezoporus occidentalis)* is one of the world's most enigmatic birds. In 1844, an expedition led by Charles Sturt departed Adelaide and headed north in search of Australia's fabled inland sea. The expedition was forced to return to Adelaide without attaining its objective, but not before one of its member shot what was initially thought to be a Ground Parrot *(P. wallicus)*. This specimen turned out to be the first Night Parrot collected by Europeans, a species eventually described by John Gould in 1861 from an individual collected several years earlier in Western Australia. Over the next few decades, a further 20 specimens were collected, nearly all from South Australia and most at the hand of Fredrick Andrews. A single bird was collected in 1912, but no further confirmed sightings were made until 1990, when the carcass of a female was found on the roadside in western Queensland. This fortuitous discovery was made when biologists from the Australian Museum pulled over to the side of the road to do some bird watching and one of them noticed the body next to his foot. More recently, a decapitated body of a juvenile was found by the roadside in Diamantina National Park, less than 200 kilometers from the 1990 record.

There is debate as to how Fredrick Andrews collected so many Night Parrots relative to others who have searched for the bird. It's been argued his success stemmed from the discovery of an area supporting an unusually large number of birds, the result of boom seasons or an influx of birds from regions further inland. Penny Olsen notes it may simply have been a case of Andrews collecting in an area where human impact was yet to render the habitat unsuitable for Night Parrots. The presence of high numbers of birds no doubt assisted Andrews in gaining an appreciation for the bird's nocturnal habits and its reliance on waterpoints. He noted that birds spent the day holed up in clumps of spinifex or patches of samphire, rousing themselves after dark and flying to the nearest water to drink. Birds spent the night feeding on the seeds of spinifex and chenopods, periodically pausing to drink. Since the time of Andrews, chance encounters have confirmed the nocturnal habits of the bird. Stephen Garnett and coauthors reported on a number of sightings in northwestern Queensland during 1992–1993. These sightings were made at different times throughout the night, and included observation of birds feeding on the ground.

The other famous nocturnal parrot is New Zealand's Kakapo *(Strigops habroptila)*. Like the Night Parrot, it spends the day on the ground in dense cover. Kakapo also roost in caves, in hollow logs, and under rock overhangs. They become active about an hour after sunset and return to the roost before sunrise. Breeding females may extend their foraging into daylight hours if food is limited or they have large nestlings. Kakapo have a

"Portrait of John Gould, Ornithologist" by Thomas Herbert Maguire (1821–1895), National Library of Australia.

number of adaptations related to their nocturnal habits. Bristles near the eyes and beak may assist in detecting obstructions when moving about in the dark. Observations of foraging birds suggest they use smell to locate food, and this has been confirmed by simple experiments. Kakapo have a large olfactory bulb ratio (longest diameter of olfactory bulb/longest diameter of the brain) relative to other birds outside the passerines (perching birds), suggesting they have a keen sense of smell. The Kakapo has facial disks reminiscent of those found on owls, and in the nineteenth century they were sometimes referred to as Owl Parrots. The facial disks of owls assist them in detecting prey by directing sound toward their ears. The facial disks of Kakapo may serve a similar purpose, though there is no evidence for Kakapo having exceptional hearing. We do not know if Night Parrots have adaptations to assist them to find food and move about in the dark. The examination of museum specimens does not provide any clues in this regard.

The vast majority of parrots are active during the day. Why have Night Parrots and Kakapo pursued a nocturnal lifestyle? The main benefit is avoiding diurnal predators. Until the arrival of humans, New Zealand was characterized by an absence of predatory mammals. The principal predators of parrots were diurnal raptors. Kakapo are too large to be important prey for New Zealand Falcons *(Falco novaeseelandiae)*, though three extinct raptors would have taken birds the size of Kakapo (Haast's Eagle *Harpagornis moorei*, Eyles's Harrier *Circus eylesi*, and Laughing Owl *Sceloglaux albifacies)*. By minimizing daytime activity, Kakapo reduced their exposure to these predators. A necessary part of this strategy is concealment during daylight, and the moss-green plumage of the Kakapo provides excellent camouflage when birds roost in open situations. Unlike New Zealand, the Australian environment supported a number of mammalian predators. Carnivores such as the Thylacine *(Thylacinus cynocephalus)*, and later the Dingo *(Canis lupus)*, occurred at low density and hunted large mammals. The Tasmanian Devil *(Sarcophilus harrisii)*, a scavenger and nocturnal hunter of medium-sized mammals, was absent from arid regions. The Western Quoll *(Dasyurus geoffroii)* posed the greatest mammalian threat to a nocturnal parrot in desert habitats. Overall, the pressure exerted by nocturnal predators may have been less than that applied by Australia's diurnal raptors. Being active at night also limits exposure to higher daytime temperatures and dehydration, and may reduce competition with diurnal parrots for food. A nocturnal lifestyle may allow Night Parrots to stay on in arid areas when seasonal conditions force diurnal parrots to relocate.

Are parrots good for the environment?

Parrots help prevent any single tree species dominating a forest and ensure individuals within a tree species have a regular rather than clumped distribution. They do this by consuming seeds on the plant before they have a chance to be dispersed by the wind or other animals. There is little information on the level of damage parrots cause to seed crops. In the Neotropics, a number of studies suggest that parrots remove around 10–20% of the seed crop of particular tree species. This figure can be higher where habitat has been fragmented, and parrots are forced to rely on fewer resources. The level of seed destruction only has ecological meaning if such losses influence the distribution and abundance of plants. Only a handful of studies have considered the role of parrot seed predation in seed dispersal. Most of these dealt with a single tree species and focused on the relationship between the tree and its primary seed disperser. Further work in this area could be profitably undertaken on groups of Pacific islands supporting similar ecosystems, but differing in the composition of their seed predator and seed disperser communities.

Henry Howe looked at seed production, dispersal, and predation in the Panamanian rainforest tree *Tetragastris panamensis*. He found that Red-lored Parrots *(Amazona autumnalis)* and Mealy Amazons *(A. farinosa)* destroyed around 6% of the seed crop, but these losses were only a small fraction of the >70% of seeds that were effectively wasted (i.e., not removed from the tree by seed dispersers). There was also considerable waste among dispersed seeds, with most seedlings failing to establish due to competition within fecal clumps. In Tonga, Red Shining-parrots *(Prosopeia tabuensis)* destroy around 16% of the *Myristica hypargyraea* seed crop, an amount unlikely to negatively impact on *Myristica* regeneration strategies. This large-seeded tree is dispersed by the Pacific Imperial-pigeon *(Ducula pacifica)*. Seeds consumed by pigeons are deposited under other *Myristica* trees, which assists in the maintenance of genetic diversity. Seed dispersed in this way has a 6.5% chance of becoming established. Pigeons consume around half of the *Myristica* seed crop, with a quarter of the crop falling to the ground.

Parrots are not usually considered seed dispersers, targeting the seed rather than any associated food reward (such as the pulp surrounding the seed). Typically, they process the seed before consumption, removing the possibility of viable seed passing through the digestive tract. Nevertheless, there is the potential for seed dispersal when parrots forage on fruit pulp or other food rewards. Theodore Fleming and coworkers fed ripe fruit of the Neotropical pioneer tree *Muntingia calabura* to captive Orange-chinned Parakeets *(Brotogeris jugularis)*. The birds sucked pulp and seed from the fruit, but no manipulation of seed was observed. Seeds collected from the bird's droppings were able to be successfully germinated. They concluded that parakeets were a major seed disperser of *M. calabura* within their study area. In Madagascar, Black Parrots *(Coracopsis nigra)* are the primary seed disperser of the dominant canopy tree *Commiphora guillaumini*. The black seeds of this tree are partially enclosed in a colorful and nutritious food reward, which the parrots nibble off before dropping the seed. Black Parrots sometimes fly off with seeds in their beaks, ensuring that some seed is dispersed beyond the tree crown. These dispersed seeds are much more likely to establish themselves as seedlings than those falling beneath the crown.

Nectar is commonly reported in the diet of parrots and can be especially important when other foods are in short supply. Parrots with general diets that feed on nectar are often robbing the plant of food intended for pollinators such as hummingbirds or insects. Visitation by such parrots can result in widespread destruction of flowers, with potentially negative consequences for seed production. However, some plants seek to attract parrots and rely on them wholly or partially for pollination. In these cases, the parrots have physical or behavioral adaptations for nectar feeding and tend not to harm the plant.

The flowers of the Amazonian canopy tree *Moronobea coccinea* are well adapted for pollination by parrots. They are colorful, robust, and arranged vertically on horizontal branches. Golden-winged Parakeets *(Brotogeris chrysoptera)* have been observed visiting these flowers to feed on nectar and pollen, their beaks becoming covered in pollen in the process. Flower morphology ensures that pollen-laden beaks inserted into flowers contact the flower's stigmas. Cross-pollination is facilitated by birds visiting multiple flowers on a tree, while the opening of only a small number of flowers each day encourages birds to visit multiple trees. In the Colombian Amazon, several species of parrots feed on the nectar and flowers of *Erythrina fusca*, though only Dusky-headed Parakeets *(Aratinga weddellii)* and Cobalt-winged Parakeets *(Brotogeris cyanoptera)* provide pollination services. The brightly colored flowers are enclosed to help prevent the theft of nectar and pollen by insects and some birds. Dusky-headed Parrots use their bills to open the flowers without causing damage. They then reach in to drink the nectar, their chin and throat receiving a dusting of pollen. Cobalt-winged Parrots employ a slightly different technique, which results in pollen being deposited on their foreheads.

In Tasmania, successful reproduction in Swift Parrots *(Lathamus discolor)* is dependent on the availability of nectar and pollen from the flowers of Tasmanian Blue Gums *(Eucalyptus globulus)*. Feeding parrots transfer pollen from anthers to stigmas, assisting with pollination. Swift Parrots are not the only pollinators of Tasmanian Blue Gums, with lorikeets, honeyeaters, and insects also contributing. Swift Parrots, and probably lorikeets, are the most effective pollinators of Tasmanian Blue Gums for a number of reasons. The way Swift Parrots feed ensures that they accumulate large quantities of pollen on areas likely to contact stigmas, such as the beak, chin, and tongue. The smaller, finer bills of honeyeaters carry less pollen and are less likely to contact stigmas. Seed production in individual Tasmanian Blue Gums is dependent on the availability of pollen from other trees. Insects are less likely to move between trees than birds and are thus less effective pollinators. Variable seed production at different heights in the canopy provides evidence of the effectiveness of birds versus insects as pollinators. Limited seed production occurs low in the canopy, an area typically avoided by foraging birds. The decline in the Swift Parrot population or the displacement of birds from flowers by introduced honeybees means there is potential for seed production in Tasmanian Blue Gum populations to be negatively affected.

Parrots are messy feeders, and the ground beneath fruiting trees is littered with debris. You often find fruit and intact seeds on the ground, dropped or dislodged by foraging birds. Dropped seeds may germinate where they fall or be dispersed by terrestrial animals. The benefits of such

Swift Parrot *(Lathamus discolor)* foraging on eucalypt flowers. Chris Tzaros

accidental dispersal depend on the ecology of the food plant. In fire-prone ecosystems, many plants have adopted a strategy of storing seed in the canopy and releasing it en masse following fire. There is a greater likelihood of germination and seedling establishment after the fire due to reduced competition for resources and the swamping of seed predators. Enclosing seeds in hard woody fruits protects them from high fire temperatures and helps deter seed predators. One downside of this strategy is that populations of plants may fail to regenerate in the absence of fire. Some spontaneous release of old seed occurs, but this may be less viable than younger seed. The dispersal of small quantities of viable seed during the intervals between fires provides insurance against long fire intervals. Parrots are one of the few seed predators capable of assisting in this regard, a combination of morphology and cognitive ability allowing them to process woody fruits.

Glossy Cockatoos *(Calyptorhynchus lathami)* in central New South Wales, Australia, feed on two species of shrubby sheoak, both of which store seed in the canopy and regenerate after fires. Glossy Cockatoos use their specially adapted bill to extract and process the seed from the woody sheoak cones. In the process, they drop cones or discard remnants that contain intact seed. Their foraging activities also result in damage to branches, causing cones to dry out and release their seed. This light seed rain contributes to the partial regeneration of stands between fires and helps prevent stands going extinct when the interval between fires exceeds the lifespan of individual plants. Occasionally, a Glossy Cockatoo will fly off clutching a cone it was about to feed on. Such events are rare opportunities for long-distance dispersal in sheoaks. That sheoaks gain some benefit from the depredations of Glossy Cockatoos is suggested by the bright orange color of cone fragments that litter the ground beneath feed trees. These are highly visible to overflying cockatoos and may serve to attract birds to profitable feeding patches. Given the potential for Glossy Cockatoos to strip favored feeding trees of cones, it seems improbable that a predator-attracting trait would persist unless the plant gained some benefit. Whether or not interactions between Glossy Cockatoos and sheoaks are beneficial to both parties remains to be investigated.

Parrots themselves are food for a range of predators, being especially important to birds of prey. Tom Aumann studied raptor diets in the unpredictable climate of central Australia. The importance of parrots varied between species, dominating the diet of bird eaters such as the Grey Falcon *(Falco hypoleucos)*. When they were abundant, Budgerigars *(Melopsittacus undulatus)* were important to a wide range of raptors. Prey remains or pellets collected from raptor nests by Tom Aumann showed that for some pairs in some years, Budgerigars comprised between 40–86% of the diet. A reliance on parrots means there is the potential for raptor populations to fluctuate in response to variations in parrot numbers. Within Tom Aumann's study area, the number of occupied territories, the number of breeding pairs, and the number of young produced per nest all increased when food was abundant. The extent to which the size and distribution of parrot populations are limited by raptors is less clear. Small populations of parrots can be vulnerable when targeted by raptors, especially when the hunters are sustained by other prey species. Island parrot populations are susceptible to introduced predators, and declines in New Zealand parrot populations have been linked to predation by stoats, rats, and possums.

Is it true that some moths depend on parrots?

In Australia, moths belonging to the genus *Trisyntopa* complete their life cycle within the confines of parrot nests. One species is associated with Eastern Rosellas *(Platycercus eximius)* and Mulga Parrots *(Psephotus varius)*, though it may also occupy the nests of other Australian parrots. The remaining two species in the genus are each associated with parrot species that nest in terrestrial termite mounds. *Trisyntopa scatophaga* occurs with Golden-shouldered Parrots *(Psephotus chrysopterygius)* and the recently described *Trisyntopa neossophila* with Hooded Parrots *(Psephotus dissimilis)*.

Antbed parrots excavate nests in terrestrial termite mounds. Moths lay their eggs in these nests, the larvae emerging when the parrot eggs hatch. This ensures that larvae have a ready supply of food in the form of nestling excrement. Once the parrot nestlings leave the nest, the larvae move to the chamber walls and pupate. The adults emerge the following season, and the cycle is repeated. The antbed parrot moths are dependent on the parrots, but are the parrots dependent on the moths? A clean nest is less likely to become infested with parasites or attract predators, factors which should contribute to greater nest success. Studying Hooded Parrots, Stuart Cooney was unable to detect any difference in nestling growth rates and reproductive success between nests having the moth and those from which moths had been experimentally removed. On the basis of this study, conducted over a single breeding season, the moth-parrot relationship appears to be one-sided. The moth benefits, while the parrot gains no measurable advantage or disadvantage.

There is an important footnote to this tale of moth-parrot relationships. Australia once had three antbed parrots, with the Paradise Parrot *(P. pulcherrimus)* last sighted in the 1920s and now presumed extinct. It is likely the Paradise Parrot had its own unique moth, the extinction of which coincided with the disappearance of Paradise Parrots from the downs country of southeast Queensland. The future of the remaining antbed parrots in the wild is by no means secure. The Golden-shouldered Parrot is endangered and the Hooded Parrot has suffered a contraction in range. In their description of *Trisyntopa neossophila*, Ted Edwards and coauthors persuasively argue that introducing moths into captive populations of antbed parrots may provide insurance against extinction of the moth in the wild. Further investigations of the moth–parrot relationship may reveal that antbed parrots derive some benefit from the moth. If this proves to be the case, the success of any future antbed parrot reintroductions may depend on the concurrent release of their moth partner.

Chapter 6

Reproduction and Development

Where do parrots nest?

Most parrots nest in holes in trees, relying on natural cavities or those excavated by other species. Nests are often high in the canopy, posing a challenge for parrot biologists. Single rope techniques employed by cavers are used to access nests. A slingshot is employed to fire a line over a branch above the nest. A climbing rope is attached to this line and hauled into the canopy and back to ground level, where it is fastened to a nearby tree. The free end of the rope is climbed using mechanical ascenders that slide up the rope, but lock into position when weight is applied to them. With proper training, single rope techniques provide a safe and secure method of gaining access to the canopy. Early naturalists and egg collectors did not have it so good. At Iron Range National Park on Cape York, there is a large old fig known as the Smugglers Tree. Protruding from its trunk is a line of metal spikes once used by poachers to climb to the nests of Eclectus Parrots *(Eclectus roratus)*. Nerves of steel would have been required to ascend to heights >30 meters by this means. I have observed similar spikes, rusting and mostly buried by the expanding girth of the tree, sticking out from the trunks of eucalypts in the Western Australian wheatbelt. These trees are still used by Carnaby's Cockatoo *(Calyptorhynchus latirostris)* for nesting.

Tree hollows have a number of advantages as nest sites. They provide protection from predators and shelter from the weather. The microclimate within cavities is relatively stable, buffering birds from extremes of temperature and humidity. These benefits are maximized in deep hollows with small entrances. Accordingly, parrots choose the smallest hollow that can accommodate themselves and their young. If a hollow is too small, birds may widen the entrance or further excavate the cavity. Denis Saunders and fellow researchers reported on hollow use by parrots in the Western Australian wheatbelt. They found entrance diameter, internal diameter, and hollow depth all increased with increasing parrot size, though there was overlap between species. The number of hollows at a given locality is fixed in the short-to-medium term, and parrots may have to compete among themselves and with other species for access to nest sites. A shortage of hollows may limit the size of populations.

Natural tree cavities are formed when damage to sapwood allows decay-causing organisms access to the vulnerable heartwood. The actions of fungi and insects result in wood decay, which is exposed when branches break off or the trunk snaps. Wind storms and fire play a role in the initial wounding of trees and the subsequent exposure of decayed wood. The actual hollow used by parrots for nesting is created when this exposed material is subsequently eroded or excavated. Parrots also use tree hollows created by woodpeckers, though this family of birds is absent from Australasia. A small number of parrots excavate their own hollow. On Cape York, Double-eyed Fig-parrots *(Cyclopsitta diophthalma)* excavate nests in dead trees or the dead branches of living trees. In New Guinea, Pesquet's Parrots *(Psittrichas fulgidus)* excavate hollows into the trunks of dead trees, preferring species with soft wood. Observations of nesting *Geoffroyus* par-

Red-tailed Cockatoo *(Calyptorhynchus banksii naso)* at entrance to her nest hollow, southwest Western Australia. Tony Kirkby

rots in New Guinea and the Solomon Islands suggest they excavate their own nest hollow into the trunks of dead or damaged trees.

Quiet a few parrots use nest sites other than tree hollows, including termitaria, cliffs, and burrows. Use of tree hollows is the ancestral condition, and many species that use alternate nest sites can still use tree hollows. The shift to alternate sites is thought to be a response to intense competition or high levels of predation associated with use of tree hollows. Donald Brightsmith could find no evidence of competition between small cavity nesters in a pristine Neotropical forest. He did find evidence for higher predation rates of nests located in tree hollows relative to those in arboreal termitaria, concluding that predation was the most likely reason birds had shifted nesting substrate. This conclusion is supported by the shorter nestling periods of tree hollow nesters compared to those using alternate sites. High predation rates should select for shorter nestling periods as this reduces the time nestlings are exposed to predators. Nevertheless, competition for nest sites may be a factor in the shift from tree hollows to alternate sites in open habitats where hollows are less abundant.

Many parrots excavate their nest into arboreal termitaria. Donald Brightsmith reported on the use of arboreal termitaria by *Brotogeris* parakeets in the Peruvian Amazon. The termitaria used by parrots were globular structures having a volume of around 100 liters and attached to trees or palms at a height of approximately 10 meters. Cobalt-winged Parakeets *(Brotogeris cyanoptera)* excavated long narrow tunnels that sloped upward and discharged into the base of the nesting chamber. Entrance tunnels had a diameter of 4.8 centimeters and a length of 22 centimeters, while nesting chambers were 17 centimeters in diameter and 16 centimeters high. Active termite mounds were preferred, possibly because they were less prone to collapse. Parrots also chose to excavate nests in termitaria containing *Dolichoderus* ants. These ants don't have a nasty sting, but readily swarm when disturbed. This behavior may be sufficient to deter predators. They may also play a role in nest sanitation or otherwise obscure the presence of nesting birds. Horace Webb provided details on cavities excavated by Finsch's Pygmy-parrot *(Micropsitta finschii)* into arboreal termitaria on Santa Isabel, Solomon Islands. Entrance holes averaged 3.3 centimeters in diameter and were located on the underside of mounds. Entrance tunnels curved upward, opening into the top of nesting chambers approximately 10 centimeters high and 15 centimeters long. The excavated cavities were sealed off from the rest of the mound by the host termites.

There is a group of three Australian grass parrots that excavate their nests in large terrestrial termite mounds. This allows them to occupy lightly timbered grasslands, habitat that provides abundant food and few competitors in the form of tree-nesting parrots. Stuart Cooney studied the Hooded

Parrot *(Psephotus dissimilis)* near Katherine in the Northern Territory. In his study area, two species of termite produced different shaped mounds. Birds preferred to nest in the cathedral-like mounds of the Spinifex Termite *(Nasutitermes triodiae)*, possibly because of their greater thermal mass. Parrots nested in areas supporting higher densities of these mounds. Birds avoided inactive mounds and those with low levels of activity. Nest success was positively correlated with termite activity. The presence of termites is necessary for maintaining an equable microclimate within the parrot nest cavity. High levels of termite activity may reduce predation on parrot nests, either by acting as a physical deterrent or masking the presence of nesting parrots. Given that not all termite mounds are suitable for nesting, there is potential for anthill parrot populations to be limited by nest site availability. Land management activities that reduce the natural densities of termite mounds may negatively impact on parrot populations.

In a number of tree-nesting parrots, birds prefer to nest close to one another. Regent Parrots *(Polytelis anthopeplus)* may nest in colonies containing as many as 18 pairs, with up to four nests located in a single tree. The nests of Thick-billed Parrots *(Rhynchopsitta pachyrhyncha)* are often aggregated, with up to 31 nests recorded in a single cluster and as many as three nests in a single tree. Other species nest in looser coalitions, with nests sometimes separated by as much as one kilometer. While social or ecological factors may explain such behavior, it is only possible where relatively high densities of tree hollows occur in suitable breeding habitat. This is dependent on the age and history of tree stands or the density of primary excavators such as woodpeckers. As suggested by Jessica Eberhard, the reliance parrots have on preexisting cavities may explain the low incidence of colonial breeding (birds nesting in close proximity at the same time). A number of colonial breeding parrots have gotten around limitations placed on breeding density by the abundance of tree hollows or termitaria. They achieved this by excavating nests into cliffs or building them from sticks.

Burrowing Parrots *(Cyanoliseus patagonus)* excavate their burrows into sandstone, limestone, or earth cliffs. The largest Burrowing Parrot colony is located in northeastern Patagonia, where more than 50,000 burrows have been dug into soft sandstone along 12.5 kilometers of ocean cliffs. Burrows are around 150 centimeters long, 26 centimeters wide, and 13 centimeters high. Juan F. Masello and colleagues found this colony to have the highest nestling survival rates of any parrot, which they attributed to an absence of predation. Terrestrial predators have difficulty climbing the unstable cliffs, while the large number of parrots in the area deters avian predators. Monk Parakeets *(Myiopsitta monachus)* build large stick nests in trees or on cliffs. These contain a number of chambers, each used by a different pair or group of birds. Colonies are formed when nests are built in

Large numbers of birds and steep cliffs combine to deter nest predators at the world's largest Burrowing Parrot *(Cyanoliseus patagonus)* colony. Petra Quillfeldt

close proximity. Most breeding occurs within colonies, possibly due to the increased likelihood of females sitting on eggs (incubating) being alerted to the approach of predators.

Some species of African lovebirds build their own nests, though these are constructed within tree hollows, rock crevices, or the communal nests of weavers. Rosy-faced Lovebirds *(Agapornis roseicollis)* establish small colonies within the large grass nests of Sociable Weavers *(Philetairus socius)*. Birds cut branches and strip bark from trees for use as nesting material, which is then tucked under the feathers of the rump and flanks for transport back to the nest. Hanging-parrots *(Loriculus)*, a group of birds closely related to the lovebirds, have also been reported as building their own nests. Francine Buckley reported on nest-building behavior in Blue-crowned Hanging-parrots *(Loriculus galgulus)* and Vernal Hanging-parrots *(L. vernalis)*. Birds were taken from the wild and observed in aviaries. Male and female Blue-crowned Hanging-parrots cut strips from provided newspaper, with a few females tucking short strips into their throat and breast feathers. Vernal Hanging-parrots tucked five centimeter strips of paper into their rump and upper tail coverts, a carrying technique similar to that employed by lovebirds. On Cape York, Palm Cockatoos *(Probosciger aterrimus)* construct a platform of sticks within tree hollows. These may be suspended or rest on decayed material within the tree trunk. Observations by Stephen Murphy suggest the construction of suspended platforms commences with birds dropping large sticks into the hollow, some of which lodge in the cavity and serve as a foundation for the remaining structure. A fine layer of splinters is added to the platform immediately prior to laying eggs.

Do parrots nest at the same time every year?

Breeding in parrots is timed so that periods of peak food demand coincide with periods of food abundance. Peak demand occurs midway through the nestling period, requiring birds to commence breeding before food becomes abundant. They use environmental cues, such as day length and temperature, to get the timing right. In temperate Australia, most parrots commence breeding in late winter or early spring, continuing into summer if conditions remain favorable. Adequate soil moisture and increasing temperatures result in high levels of plant growth at this time, leading eventually to plant reproduction and the setting of seed. There are anomalies to this overall pattern. The breeding seasons of dietary specialists are closely tied to the reproductive cycles of their preferred foods. In central New

Adult female Grey-headed Lovebird *(Agapornis canus)* and nestlings.

Heinz Lambert

South Wales, the majority of Glossy Cockatoo *(Calyptorhynchus lathami)* eggs are laid in autumn. Breeding in Glossy Cockatoos depends on the availability of young *Allocasuarina* cones. These are produced the previous winter, with cones maturing and becoming available to birds around the middle of summer. Lorikeets on the inland slopes of New South Wales have young in the nest over winter, reflecting the availability of pollen and nectar from winter-flowering eucalypts.

In tropical Australia and islands further north, different parrots nest at different times. Food availability appears to be the main influence on breeding season, though avoiding hollow-flooding rains may be a factor. Hooded Parrots *(Psephotus dissimilis)* nest during the wet season, with breeding commencing in January and completed by the end of April. Peak food availability for this species occurs around the end of the wet season, coinciding with young birds departing the nest (fledging). On Cape York, the majority of Eclectus Parrots *(Eclectus roratus)* begin nesting in the dry season, with young fledging during the wet season when food is likely to be more abundant. If rain continues into the dry season, breeding may be delayed as birds wait for nest hollows to dry out. Palm Cockatoos *(Probosciger aterrimus)* on Cape York follow a different pattern, with birds at various stages of the nesting cycle encountered in all months. The building of a stick platform inside the nest hollow allows them to breed throughout the wet season without fear of flooding. At Crater Mountain in Papua New Guinea, high rainfall throughout the year means one fruit or another is always in season. The broad dietary preferences of Eclectus Parrots allow them to take advantage of this situation, with breeding reported in most months of the year. Palm Cockatoos at Crater Mountain have more limited diets, relying on a few species of fruit. As a consequence, they have a distinct breeding season coinciding with increased fruiting in food trees.

Donald Brightsmith has studied Psittaciform nesting habits at the Tambopata Research Center in southeastern Peru. The surrounding lowland rainforest contains up to 20 species of parrot, many of which are relatively abundant. Annual rainfall is 3,200 millimeters, with a distinct dry season from April to October. Smaller parrots breed in the dry season, while larger parrots breed during the wet season. The different diets of the two groups may explain the observed patterns. Small parrots eat more nectar, flowers, and small seeds than large parrots, possibly because these foods are too energetically costly for the latter to harvest efficiently. Nectar, flowers, and small seeds are more abundant in the dry season, while fleshy fruit production peaks early in the wet season.

In the Caribbean, Amazon parrots typically lay their eggs during the dry season. The timing of egg laying in Puerto Rican Amazons *(Amazona vittata)* is thought to be related to the fruiting of major foods, though

Female Eclectus Parrot *(Eclectus roratus)* nestling retrieved from its nest for measurement, Cape York, Australia. Matt Cameron

Scarlet Macaw *(Ara macao)* nestlings are lowered to the ground for measurement, Tambopata Research Center, Peru. Matt Cameron

the availability of dry nesting cavities is an equally plausible explanation. Lilac-crowned Amazons *(A. finschi)* and Yellow-crowned Amazons *(A. ochrocephala)* inhabiting dry tropical forests in western Central America also breed during the dry season. Katherine Renton monitored changes in the abundance and diversity of Lilac-crowned Amazon food resources over the course of a year. Food abundance peaked late in the wet season prior to eggs being laid, with a second peak in the early to mid dry season when parrots were raising young.

Long nestling periods and extended post-fledging care of young mean breeding twice a year is not an option for most large parrots. Even if a second nesting attempt could be accommodated, the potential negative impacts on adult survival make it a high-risk option. On Cape York, a small number of Eclectus Parrot females attempt to nest again in the same season after successfully fledging a first brood, with around half of these attempts being successful. Smaller parrots are less constrained, and many species breed twice in one season. Long-term observations by John Courtney of Little Lorikeets *(Glossopsitta pusilla)* and Musk Lorikeets *(G. concinna)* on the northwest slopes of New South Wales have shown both species will breed twice a year when critical food resources are abundant. James Waltman and Steven Beissinger found two-thirds of Green-rumped Parrotlets *(Forpus passerinus)* pairs that bred successfully early in the season nested again later in the season. Nests commenced late in the season produced just as many young as those from early in the season.

Do parrots use the same nest every year?

The level of cavity reuse by parrots varies between species and locations. When looking at the results from field investigations, it is important to take into account the length of the study. Long-term studies tend to report higher levels of reuse, with cavities remaining unoccupied for a number of years before reoccupation. Studies of Amazon parrots have shown relatively high levels of cavity reuse. A five-year study of Blue-fronted Amazons *(Amazona aestiva)* reported 62% of nest hollows were reused, while a 14-year study of Lilac-crowned Amazons *(A. finschi)* showed 42% of nest hollows were used at least twice. In marked populations it is possible to determine if cavities are being reused by the same or different birds. In the aforementioned study of Blue-fronted Amazons, researchers captured and banded nesting females. They found no evidence that females switched hollows, and most females nested in the same hollow in consecutive seasons. Burrowing Parrots *(Cyanoliseus patagonus)* return to the same burrow each year, with pairs enlarging and extending the nest cavity. It appears that high levels of cavity reuse are associated with ongoing use by the same pair of birds.

In Blue-fronted Amazons and Lilac-crowned Amazons, the probability of a hollow being reused is greater if it has previously supported a successful nesting attempt. Igor Berkunsky and Juan Reboreda were able to take this analysis one step further, demonstrating that nest success in Blue-fronted Amazons could be explained by nest hollow characteristics. They found that nests in deep hollows with thick walls were more likely to be successful. Despite the inherent logic in these findings, they have not been replicated in many other studies. Long-term investigations of the breeding biology of marked populations of cockatoos in the Western Australian wheatbelt could find no link between nest success and hollow reuse. Pairs of Pink Cockatoos *(Lophochroa leadbeateri)* that nested successfully were just as likely to shift hollows as stay put the following year. For the most part, these studies were also unable to find a relationship between nest hollow characteristics and nest success. An exception was the vulnerability of certain hollows to flooding.

How can high or low rates of cavity reuse be explained in the absence of any apparent variability in quality of nest cavities? One explanation is competition within species (intraspecific competition) or between species (interspecific competition) for nest hollows. High rates of cavity reoccupation could be expected if nest cavities were in short supply. Determining the availability of potential nest sites is problematic. Not only is it difficult to ascertain the suitability of cavities as nest sites, but apparently suitable hollows may not be available due to the territorial or ranging behavior of other hollow users. While difficult to observe, conflict between individuals is the best indicator a shortage of suitable nest hollows exists. In Cape York rainforests, Eclectus Parrots *(Eclectus roratus)* and Sulphur-crested Cockatoos *(Cacatua galerita)* frequently clash over hollow ownership. Sulphur-crested Cockatoos are often successful in temporarily or permanently taking over Eclectus Parrot nest hollows, causing delays in Eclectus Parrot breeding and lowering reproductive output. Eclectus Parrots nest in emergent rainforest trees, and a shortage of such sites means females exhibit extreme attachment to nest sites and defend them resolutely against other females.

Nest hollows in which parrots have successfully bred may go unused for two or three years, even in populations that display high levels of cavity reuse. A number of explanations have been put forward to explain this relatively common finding. One disadvantage of cavity nesting is the potential for nests to become infested with parasites or to harbor disease. Ectoparasites (external parasites such as mites) were shown to be an important cause of nest failure in Thick-billed Parrots *(Rhynchopsitta pachyrhyncha)*, with nestlings from heavily parasitized nests suffering anemia. Regularly shifting hollows or the temporary abandonment of infested hollows may prevent or control parasite outbreaks. Moving your nest from one hollow to another may thwart predators intent on returning to a site where they enjoyed easy pickings the previous season. This strategy may explain why parrots that have lost a clutch to predators will sometimes lay a replacement clutch in a new hollow. Known nest hollows not being used in a given year may still be defended by their owners, who may not be reproductively active that season or are breeding nearby in another hollow.

How do parrots reproduce?

Reproduction in parrots is straightforward. In most species, breeding is undertaken by stable mated pairs that remain together throughout the year. This renders unnecessary the elaborate courtship displays seen in birds that establish or renew pair bonds each breeding season. The female has a passive role in the courtship displays of most parrots, though in a few species she's an active participant. Courtship involves actions such as strutting, bowing, neck-stretching, head-bobbing, bill-clasping, tail-fanning, and the arching of wings. Some postures appear designed to display particular color patches. Fluttering flights are common, as is eye-blazing, the latter produced by constricting the pupil to expand the exposed outer edge of the iris. The regurgitation of food by the male to the female has been incorporated into the displays of many species, principally those where the female does all the incubation and is dependent on the male for food early in the nesting period. Head-bobbing and bill-clasping are thought to be ritualized forms of such courtship feeding. Courtship can be a noisy affair, and some calls are only made in this context.

Palm Cockatoos *(Probosciger aterrimus)* on Cape York have a spectacular display, which probably serves for courtship and territorial advertisement. Performed by males, the display was first described in 1984 by G. A. Wood:

> Early one morning I awoke to the sound of loud knocking coming from the display site near my camp. On investigation I found a Palm Cockatoo pirouetting around the top of the display tree trunk, wings outstretched, beating the trunk with an object in its left foot, as it turned. This display continued for a minute or more and was followed by head rolling and erection of the crest.

Displays also involve upside-down swinging and much calling. Woody fruits or specially fashioned sticks are used as drumming implements, a rare example of animals using tools for a purpose other than foraging. It has been suggested that drumming signals the durability of the nest hollow or the cognitive ability of the drummer. Why Palm Cockatoos on Cape York have evolved such an elaborate display compared to other parrots remains a mystery. Intense competition among males for ownership of the best nest sites and access to quality females may be involved. Information on changes in hollow ownership, paternity of nestlings, and mate swapping by females would assist in clarifying the issue.

Successful parrot courtships lead to copulation. As with most birds, male parrots do not possess a penis. Transfer of sperm from the male to the female is achieved via cloacal contact, the cloaca providing a single opening to the exterior for the digestive, urinary and reproductive systems. Jessica

Eberhard published a detailed description of copulatory behavior in wild Monk Parakeets *(Myiopsitta monachus)*. Typical of Neotropical parrots, the male does not fly onto the back of the female. Rather, he places one foot on her back, retaining a grip on his perch with the other. Bringing his tail under the females, he presses his cloaca against hers and waggles his tail. Copulation in Monk Parakeets occurs in bouts, with males mounting females several times over a period of four to five minutes.

I have observed wild Glossy Cockatoos *(Calyptorhynchus lathami)* copulating on several occasions, usually after their evening drink. Copulations occurred on stout, rough-barked horizontal branches and were preceded by a courtship display and courtship feeding. Courting males bowed and fanned their tail, fluttered between branches, and uttered a distinctive "kwee-chuck" call. If the female was receptive, the male flew onto her back. She moved her tail to the side, allowing the male to bring their cloacae into contact. Copulation lasted around a minute, the male moving his tail up and down throughout, wings spread for balance. Initial attempts to mount the female were often rebuffed, as were further attempts once copulation had been achieved.

Copulations in Vasa Parrots *(Coracopsis vasa)* can last as long as 90 minutes. These long copulations are made possible by the presence in males of a large erectile cloacal protrusion, which serves to tie the female to the male during copulation. The closely related Black Parrot *(Coracopsis nigra)* possesses a similar structure. Jonathan Ekstrom and colleagues described long copulations in Vasa Parrots. Following insertion of the male's cloacal protrusion, tied birds sat alongside each other. After a brief interlude, during which the female was fed, the male began to make pumping movements that lasted for about half the period birds were locked together. Vasa Parrots also practice short copulations that last a few seconds. Why have long copulations evolved? Female Vasa Parrots copulate with several males, meaning that nestlings within a brood may have different fathers. Long copulations may increase the likelihood of males achieving paternity.

How many eggs do parrots lay?

Clutch size varies considerably among Psittaciformes, but is broadly consistent with other birds having helpless young. Large cockatoos produce the smallest clutches. The Palm Cockatoo *(Probosciger aterrimus)* and Glossy Cockatoo *(Calyptorhynchus lathami)* only ever lay a single egg, while Red-tailed Cockatoos *(C. banksii)* occasionally produce a second egg. Other black-cockatoos usually lay two eggs, though single egg clutches are not uncommon. Data on wild Gang-gang Cockatoos *(Callocephalon fimbriatum)* are lacking, but captive birds usually lay two eggs. Clutch size is greater in the remaining cockatoos and most other parrots. A study of Lilac-crowned Amazons *(Amazona finschi)* reported a median clutch size of three eggs, while a study on Galahs *(Eolophus roseicapillus)* in the Western Australian wheatbelt found a median clutch size of four eggs. Similar clutch sizes have been reported for other medium- to large-sized parrots. The smallest parrots are responsible for the largest clutches. Clutch size in Green-rumped Parrotlets *(Forpus passerinus)* exceeds that of many tropical birds, with an average of seven eggs and a maximum of 12 eggs.

Clutch size evolved to maximize the number of birds parents contribute to the population over the course of their lifetimes. It seems logical that larger clutches will result in more young birds, but this is not necessarily the case. Parents may have trouble finding enough food for large numbers of nestlings, meaning that some die in the nest or fledge in poor condition with limited survival prospects. Large clutches also require a lot of energy to produce and may not be able to be replaced if predators destroy the initial clutch. These facts mean small clutches can result in more birds entering the population compared to large clutches. There may be benefits in reducing clutch size still further. Raising large numbers of young birds may reduce the survival prospects of parents, with the need to forage for longer periods increasing exposure to predators and placing added wear-and-tear on the body. Producing fewer young each year over a greater number of years may be the best strategy for maximizing reproductive output over a lifetime.

While parrots are often described as having small clutches, clutch sizes of many small- to medium-sized parrots are large relative to other bird groups. This pattern is even more striking when we consider the largely tropical distribution of the Psittaciformes, with tropical birds having smaller clutches than temperate species due to the more equable climate of tropical environments and associated stability in food resources. There are a number of possible explanations for the evolution of larger clutch sizes in hollow nesters. Cavity nesters are exposed to lower predation rates than those nesting in the open, meaning the risks involved in laying large clutches are less than would otherwise be the case. If the supply of cavities is limited, it makes evolutionary sense for birds in possession of hollows to maximize their reproductive output. These factors may have contributed to the extraordinary clutch size in Green-rumped Parrotlets, which have among the highest reported breeding success of any parrot and experience intense intraspecific competition for nest sites.

The trend for increasing clutch size with decreasing body mass in parrots reflects differences in life history. Among birds generally, larger species have slower life histories. They live longer, breed later in life, take longer to raise their young, and produce fewer young each breeding season. In

Table 6.1. Breeding parameters and breeding success of wild parrot populations ranked by parrot size. Breeding success calculated as the percentage of eggs that produce a fledgling.

Species	Weight (g)	Average clutch size	Incubation period (days)	Nestling period (days)	Breeding success (%)
Green-rumped Parrotlet[1] (*Forpus passerinus*)	24–36	7 (5–10)	18–22	28–35	67
Budgerigar[2] (*Melopsittacus undulatus*)	26–29*	4.6 (2–7)	17–19	35	40
Rosy-faced Lovebird[3] (*Agapornis roseicollis*)	43–63*	4.4 (3–8)	23 (22–23)	43 (42–44)	14
Monk Parakeet[4] (*Myiopsitta monachus*)	115	5.6 (5–6)	24*	—	30
Crimson Rosella[5] (*Platycercus elegans*)	120–170	5.3 (3–8)	19.7 (16–28)	35*	50
Burrowing Parrot[6] (*Cyanoliseus patagonus*)	225–340	3.8 (2–5)	24–25*	62.7 (53–68)	82
Lilac-crowned Amazon[7] (*Amazona finschi*)	282–312	2.6 (2–4)	28	60	38
Thick-billed Parrot[8] (*Rhynchopsitta pachyrhyncha*)	300*	2.7 (1–5)	27 (25–32)	—	60
Galah[9] (*Eolophus roseicapillus*)	311	4.3 (2–8)	23.4 (22–26)	49.4 (46–59)	39
Blue-fronted Amazon[10] (*Amazona aestiva*)	400*	2.5 (1–6)	30	56	41
Eclectus Parrot[11] (*Eclectus roratus*)	500–600	1.8 (1–2)	30	86 (79–97) M 79 (72–86) F	18
Glossy Cockatoo[12] (*Calyptorhynchus lathami*)	520	1	30–31	90 (84–96)	42
Western Corella[13] (*Cacatua pastinator*)	527	2.7 (1–4)	22–25	60 (52–68)	59
Carnaby's Cockatoo[14] (*Calyptorhynchus latirostris*)	660	1.8 (1–2)	28–29	70–77	35
Palm Cockatoo[15] (*Probosciger aterrimus*)	550–1000*	1	30–32	65–79	22
Blue-and-yellow Macaw[16] (*Ara ararauna*)	907–1240	2.6 (2–3)	26–28*	90–94*	19

References: [1]Beissinger and Waltman (1991), [2]Wyndham (1981), [3]Ndithia et al. (2007), [4]Eberhard (1998), [5]Krebs (1998), [6]Masello and Quillfeldt (2002), [7]Renton and Salinas-Melgoza (2004), [8]Monterrubio et al. (2002), [9]Rowley (1990), [10]Seixas and Mourão (2002), [11]Heinsohn and Legge (2003), [12]Garnett et al. (1999), [13]Smith (1991), [14]Saunders (1982), [15]Murphy et al. (2003), [16]Brightsmith and Bravo (2006).

*Data from Collar (1997), Rowley (1997).

Female Regent Parrot *(Polytelis anthopeplus)* incubating eggs.
Victor G. Hurley

a number of parrots, clutch size is greater or smaller than expected based on body weight. Glossy Cockatoos lay a solitary egg, though are sometimes erroneously credited with the ability to lay a second. They are significantly lighter than many parrots laying larger clutches. Foraging studies have shown Glossy Cockatoos incapable of gathering sufficient food to support a second nestling. The laying of an additional egg as insurance against loss of the first may not be possible or warranted given moderate infertility rates and the ability of females to lay a replacement clutch. Lorikeets produce relatively small clutches, with island species laying two eggs and mainland species three or four eggs. The diet of birds that rely on nectar (nectarivores) and fruit (frugivores) is often low in protein. Despite behavioral and physiological adaptations designed to maximize protein intake, lorikeets may struggle to obtain sufficient protein to produce large numbers of eggs or support the growth of large numbers of nestlings. Increased competition for resources among island species may dictate smaller clutch sizes.

Long-term reproductive studies allow scientists to investigate the relationship between environmental conditions and year-to-year variation in clutch size. If the food available for breeding can be anticipated at the time of egg laying, it makes sense for clutch size to be adjusted accordingly. Female condition at the beginning of the breeding season is the most plausible mechanism, though it is possible more direct mechanisms are involved. In temperate Australia, clutch size in Crimson Rosellas *(Platycercus elegans)* and Galahs varies with rainfall. Extreme events have the greatest effect, with wet years resulting in large clutches and drought years producing small clutches. Breeding success is high during wet years, confirming the link between rainfall and food availability and emphasizing the benefits of increased clutch size under these conditions. Dry conditions result in fewer pairs breeding and few new birds entering the population, but there is often no appreciable reduction in breeding success as only experienced birds attempt to nest. On the Pacific coast of Mexico, Lilac-crowned Amazons *(Amazona finschi)* breed during the dry season. They respond to above average rainfall during the preceding rainy season by producing large clutches. However, precipitation during the rainy season is not a good predictor of future food supply. As a result, investment in large clutches is often wasted.

Clutch size in some parrots is consistent between years, irrespective of environmental conditions. In Burrowing Parrots *(Cyanoliseus patagonus)*, clutch sizes in drought years and years of average rainfall are similar. A six-year study of reproduction in Western Corellas *(Cacatua pastinator)* found no significant difference in clutch size between years, despite some years receiving twice the rainfall of others. Where clutch size is not matched to food supply in a given season, some reduction in brood size is likely to occur.

Are nestlings in a brood all the same age?

Many parrots commence incubation prior to the clutch being completed, resulting in eggs hatching at different times (asynchronous hatching). As a consequence, young at different stages of development are present in the nest, with smaller nestlings placed at a competitive disadvantage and suffering increased mortality as a result. Asynchronous hatching has traditionally been viewed as an adaptation allowing efficient brood reduction during periods of food shortage. The even distribution of food among nestlings may result in nest failure or the fledging of young with limited survival prospects. Scientists have generally been unable to demonstrate the benefits that supposedly accrue from asynchronous hatching.

Steven Beissinger and fellow researchers have been studying reproduction in Green-rumped Parrotlets *(Forpus passerinus)* in the llanos of Venezuela for more than 20 years. These small parrots commence incubation with the laying of the first egg and there can be a 14-day age difference between the oldest and youngest nestlings in large clutches. This makes them a good candidate for studying the benefits of asynchrony. Parrots often nest in hard-to-reach places, making experimental manipulation of brood and clutch size difficult. Within the study area, Green-rumped Parrotlets quickly took to nesting in PVC nest boxes (one-meter-deep tubes lined with hardware cloth) attached to fence posts. This allowed researchers to easily move eggs of known age between nests to create broods that hatched synchronously and asynchronously. They found that just as many young were fledged from synchronous as asynchronous broods. When brood size was large, synchronous broods performed better due to the mortality of small young in asynchronous broods. Small nestlings died of starvation due to the inequitable distribution of food.

Why do Green-rumped Parrotlets pursue asynchronous hatching when it results in fewer young? Several possible explanations have been put forward. Commencing incubation with the first egg means the female is on hand to see off predators or competitors, reducing the risk of nest failure. Green-rumped Parrotlets experience intense competition for nest sites, and pairs without a nest site will often destroy the eggs or kill the young of breeding pairs in an attempt to take over their nest site. Delaying incubation has also been shown to reduce the viability of parrotlet eggs, with a number of factors thought responsible. Ambient temperatures are high enough for embryo growth to commence in the absence of incubation, but insufficient for normal development. Unincubated eggs may also be prone to microbial infection.

Parents may be able to compensate for any competitive disparity in asynchronous broods by offering more food to the smaller nestlings. This aspect of parental care has been examined in Green-rumped Parrotlets by Amber Budden and Steven Beissinger and in Crimson Rosellas *(Platycercus elegans)* by Elizabeth Krebs. As in the Green-rumped Parrotlet study, the Crimson Rosella study made use of artificial nest boxes to facilitate nest monitoring. In both studies, feeding visits were videotaped using a small camera mounted on the side or lid of the box and connected to a video recorder on the ground. Analysis of the captured footage showed that males and females differed in the way they allocated food to nestlings, and that food allocation varied depending on brood size and nestling hunger. Males tended to preferentially feed larger, older nestlings, but were capable of responding to the hunger of individual nestlings. This latter behavior had the potential to benefit small young. Females usually distributed food equitably or favored smaller nestlings, compensating for the effects of hatching asynchrony on smaller young. The important role played by females in this regard may indicate they are better at assessing brood condition, or that the costs associated with compensatory behaviors are less in females compared to males.

Despite the compensatory behaviors of female Crimson Rosellas and Green-rumped Parrotlets, the potential for brood reduction remained. Crimson Rosella females changed their provisioning behavior when broods were hungry, feeding first-hatched nestlings more and last-hatched nestlings less. The pattern in Green-rumped Parrotlets of later-hatched nestlings being fed more frequently by females did not extend to the last-hatched nestling in large broods. It appears that females compensate for the effects of hatching asynchrony when brood size is optimal, but may promote brood reduction when the number of nestlings exceeds the optimum. Brood reduction may be required when food is scarce, or when a greater-than-expected number of eggs hatch. Nestling mortality is greatest early in the nestling period, suggesting the elimination of nestlings hatched from eggs laid as insurance against infertility. That asynchronous hatching has an adaptive role in brood reduction is further evidenced by the small size of last eggs in Green-rumped Parrotlets, with heavier eggs more likely to hatch and nestlings from heavier eggs more likely to fledge.

What proportion of eggs hatch?

Hatching success is defined as the proportion of all eggs within a population that hatch. It averages around 70% in parrots, but most studies report significant variations within populations between years. A percentage of all eggs laid do not hatch because they are infertile, with infertility rates of approximately 20% reported in a range of cockatoo species. A number of interrelated factors can result in fertile eggs failing to hatch. Some nest cavities are prone to flooding during heavy rain, resulting in the chilling of eggs and the death of embryos. Disputes over hollow ownership can result in eggs being broken or incubation being interrupted to the point that embryo development is impaired. Predators take a proportion of all eggs and may cause the abandonment of clutches by killing parent birds. The impact

Common Brushtail Possums *(Trichosurus vulpecula)* are predators of parrot nests and can negatively impact on threatened Australian and New Zealand parrots. Victor G. Hurley

of these factors is multiplied when food shortage requires birds to spend more time foraging and less time attending the nest.

Whether a population grows or declines will depend on its reproductive output over the long term. For a population to remain stable, pairs of breeding birds need to replace themselves over the course of their lifetimes. Long-lived species that experience low adult mortality are capable of achieving this with limited reproductive output. However, such species are vulnerable to small changes in reproductive parameters. Denis Saunders studied two populations of Carnaby's Cockatoo *(Calyptorhynchus latirostris)* in areas that had experienced different levels of clearing. One population occurred in an area where substantial native vegetation remained and food was abundant. This relatively healthy population produced 0.63 young per nest. The second population occurred in an area that had been extensively cleared for agriculture. It was characterized by reduced growth rates, lower fledging weights, and poor nest success. The 0.43 young produced per nesting attempt were insufficient to maintain the population, which eventually went extinct.

Robert Heinsohn and coauthors undertook a Population Viability Analysis (PVA) for two populations of Palm Cockatoo *(Probosciger aterrimus)*. The population on Cape York had very poor nest success, with infertility and predation resulting in only 0.11 young per nest. Their analysis suggested the population was in decline, with adult birds unlikely to live long enough to support such a low reproductive rate. The authors were unable to determine whether the current levels of nest predation were abnormal, though they noted that a suite of natural predators was involved. It's possible that habitat change has increased densities of some of these native species. The population at Crater Mountain in Papua New Guinea experienced greater nest success than the population on Cape York, but was subject to high levels of harvesting for food by local people. While more information on breeding biology and adult survival was required to strengthen the model, it appeared the current level of harvest (40% of nestlings) was unsustainable.

How fast do parrots grow?

Parrots grow slowly. Andreína Pacheco and coauthors suggest the standard parrot diet of nectar, fruit, and seed is a relatively poor one, which may impose limitations on growth rates. It is also possible, they note, that evolutionary constraints on growth rates have allowed parrots to exploit nutritionally unbalanced foods. Nestling growth follows a logistic pattern in most parrots. Weight gain is slow initially, peaks early in the nestling period, and then declines as nestlings approach their final weight. It is not uncommon for nestlings to attain 80% of their fledging weight half way through the nestling period. A slight reduction in weight may occur prior to fledging. Wing growth is typically measured as the length of the folded wing. Initial growth is slow and governed by bone development. Once the primary feathers emerge, wing growth accelerates and remains relatively constant until fledging. Cavities are relatively secure nest sites, meaning there is less pressure for young birds to depart the nest. Accordingly, parrots are well developed at fledging.

Growth rates are correlated with adult body weights, with large parrots having longer nestling periods. The young of some species leave the nest earlier or later than expected given their size. Glossy Cockatoos *(Calyptorhynchus lathami)* have the longest nestling period of any cockatoo and among the longest of any parrot. Slow growth rates in Glossy Cockatoos may be due to the parent's inability to provision at a faster rate. Glossy Cockatoos are dietary specialists, feeding on seeds extracted from the woody fruits of sheoaks. While these are a nutritious and stable food resource, gathering food is a time-consuming task. A slow growth rate may have been selected due to the reduced stress this places on adult birds, potentially increasing parent longevity and lifetime reproduction. Western Corellas *(Cacatua pastinator)* have relatively short nestling periods, the two or three young in each brood experiencing rapid growth. Historically, this species specialized on the underground parts of plants, which provided an abundant food source during the winter/spring breeding season. Inland

Pink Cockatoo *(Lophochroa leadbeateri)* nestlings, approximately 15–20 days old. Victor G. Hurley

Pink Cockatoo *(Lophochroa leadbeateri)* nestlings, approximately 30–35 days old. Victor G. Hurley

populations are thought to have moved toward the coast following breeding due to a shortage of food and water in the nesting area over summer. In this species, rapid growth rates allowed breeding to be completed prior to the onset of unfavorable conditions.

There is the potential for growth rates to vary between populations or from one year to the next, reflecting differences in environmental conditions. Weather is an important influence on plant growth and thus food availability, but extreme weather conditions can also reduce the time available for foraging. Not unexpectedly, there is a good relationship between nestling growth rates and the number of nestlings that survive to fledge. In Burrowing Parrots *(Cyanoliseus patagonus)*, drought has the effect of reducing nestling growth and survival. It also accentuates the trend evident in average years of smaller young having lower survival rates than older siblings. Katherine Renton compared growth rates in Lilac-crowned Parrots *(Amazona finschi)* in years that differed significantly in terms of available food resources. Higher growth rates were recorded in the year food was abundant. There was no difference in fledging weights between years, but attaining maximum weight earlier in the nestling period may be beneficial due to declines in food availability toward the end of the dry season. Nestling survival in this species is also influenced by food availability, with the starvation of later-hatched nestlings reported in years when food was scarce.

Do both parents care for their young?

Both parents contribute to the care of young, though male and female roles differ. In most parrots, females do the incubation and brooding (sitting on hatched young). They are dependent on males for food over this period, though they will occasionally forage in the vicinity of the nest. Males transfer food to females, who in turn feed any nestlings. Glossy Cockatoo *(Calyptorhynchus lathami)* females are fed once in the evening, while Carnaby's Cockatoo *(C. latirostris)* females are fed morning and evening. Lilac-crowned Amazon *(Amazona finschi)* males return to the nest two or three times a day to feed the female. Female Green-rumped Parrotlets *(Forpus passerinus)* are fed once every hour or two. The situation in Galahs *(Eolophus roseicapillus)* and related cockatoos is different, with males and females sharing incubation and brooding duties. In all species, young cease being brooded once they are large enough to regulate their temperature, or food demands have increased to the point that both parents are required to provision the family.

Adults forage as a pair and return to the nest together to feed nestlings. Young are more likely to be fed in the morning and evening, reflecting activity patterns of adult birds. The number of feeding visits per day varies between species, with smaller species having higher visitation rates. There is little evidence for increased feeding frequency with increased brood size or nestling age. Galah parents return to the hollow every couple of hours to feed nestlings, and this visitation rate remains constant throughout the nestling period. Amber Budden and Steven Beissinger reported that Green-rumped Parrotlets pairs visited the nest once an hour to feed nestlings and that brood size had no effect on visitation rates. This suggested food supply was limited, but other lines of evidence did not support this hypothesis. Further investigation may reveal that feeding frequency is an unreliable indicator of the quantity and quality of food provided to nestlings.

Whether or not parents spend the night in the nest cavity depends on species and nestling age. Lilac-crowned Amazon females stop spending the night in the hollow when nestlings are a few weeks old, while Glossy Cockatoo females roost within the hollow for much of the nestling period. Parents are more likely to spend the night in the nest cavity if the benefits to nestlings are high and costs to themselves are low. The nest cavity may represent a highly desirable roost site because of the protection it affords from weather and predators. In these circumstances, both parents may overnight in the nest cavity. Parrots are secretive around the nest, with smaller species tending to be more wary. The risk of nest predation may be greater in small parrots due to a larger suite of predators and higher nest visitation rates. The transfer of food between males and females often takes place away from the nest. Parents approach and depart the nest silently, and do not linger once nestlings have been fed. Nestlings play their part by avoiding begging until the parent appears at the hollow entrance.

Older nestlings climb to the hollow lip to be fed and spend increasing amounts of time at the entrance prior to fledging. Glossy Cockatoos can often be observed sitting in the sun in the week prior to fledging. If startled, they plunge back into the hollow. Fledging is often associated with a feeding visit, and some authors have reported parents actively encouraging young to leave the nest. The first flight may be lengthy and is usually escorted by the parents. Parrots are strong fliers straight out of the nest, but directional control and landing take some practice. Asynchronous hatching means broods may fledge over an extended period, requiring parents to keep track of young within and outside the nest hollow. Researchers avoid visiting nests around the time birds are due to fledge as it can result in young departing the nest early.

How long do young birds stay with their parents?

Limited post-fledging care is provided by Galahs *(Eolophus roseicapillus)* and Spectacled Parrotlets *(Forpus conspicillatus)*. In these species, asynchronously hatched young fledge at different times. To assist in keeping the family together, parents guide each new fledgling to the location of earlier fledged young. The young from multiple families may assemble in the same tree or group of trees. Young remain in these crèches while adults forage, siblings gathering together to be fed when their parents return. Over the course of a few weeks, crèches are integrated into local flocks. Spectacled Parrotlet young become independent five weeks after fledging, while Galah young are effectively abandoned by their parents eight weeks after fledging. Crèches have a number of benefits. Accompanying parents while they forage may entail significant predation risk in the open habitats favored by these species. Crèches provide a relatively safe environment in which flying skills can be developed. They also provide a mechanism for rapid integration of young birds into the local population, allowing Spectacled Parrots to breed again and facilitating ongoing defense of the nest hollow by Galahs.

Alejandro Salinas-Melgoza and Katherine Renton fitted radio transmitters on Lilac-crowned Amazon *(Amazona finschi)* nestlings, allowing them to follow their progress once they left the nest. In the three to four weeks following fledging, young from several family groups were crèched while parents foraged. Crèches were located distant from nest sites and food resources, suggesting their location had been selected to minimize predation

Adult male Red-tailed Cockatoo *(Calyptorhynchus banksii)* feeding juvenile, southwest Western Australia. Tony Kirkby

risk. Young birds then accompanied their parents and were fed by them until achieving independence approximately five months after fledging. Sibling relationships did not appear to survive the breakup of families, with siblings observed in different flocks. Independent young became increasingly mobile as a prelude to dispersal away from the natal area. Gerald Lindsey and coworkers similarly radio tracked fledgling Puerto Rican Amazons *(A. vittata)*. Juvenile birds did not venture outside their nesting valley until around two months after fledging, at which time they became integrated into adult flocks. An exception was a family that moved to a nearby valley following one of the young being killed by a hawk. Nesting valleys contained a number of nests, but the young produced from each of these did not associate with each other.

The longest periods of parental care are found in the largest parrots. Denis Saunders found that juvenile Carnaby's Cockatoo *(Calyptorhynchus latirostris)* stayed with their parents until the start of the following breeding season. Family groups of Glossy Cockatoos *(C. lathami)* are regularly encountered outside the breeding season, suggesting young birds remain with their parents for much of this period. The commencement of breeding appears to be a catalyst for the breakup of black-cockatoo families. If pairs do not breed the following season, young birds will often continue to associate with their parents. This may ensure that immature birds survive periods of environmental stress at minimal cost to their parents. Fledgling Kakapo *(Strigops habroptila)* stay with their mother for eight to nine months, with this period of dependency being similar for male and female offspring. Following independence, juvenile Kakapo continuously lose weight until they are about two years old. This weight loss reflects their inexperience and highlights the benefits of extended parental care.

Do parrots pair for life?

Parrots are said to pair for life, an assertion difficult to prove given the longevity of many species. Most studies that follow marked individuals over a number of breeding seasons find that at least some birds that bred together one season are mated to different birds or unmated the following year. A study of Pink Cockatoos *(Lophochroa leadbeateri)* that extended over several years found pairs remained together until one of them died. A study in the Western Australian wheatbelt found a quarter of Western Corella *(Cacatua pastinator)* pairs that bred together for the first time were separated the following year. The separation rate for all Western Corella pairs was 15%, with breakup recorded in one pair that had been together for at least three years. Higher rates of pair separation have been recorded in Green-rumped Parrotlets *(Forpus passerinus)*, with between one-third and two-thirds of pairs changing mates between years. The reasons why parrots seek new partners are not understood, though breeding success in the year preceding separation does not appear to be a factor.

The basic social unit of most parrots is the mated pair, with the pair bond reinforced through the sharing of food, mutual preening, coordination of activities, and vocal exchanges. Apart from allowing the cooperative care of young, stable pairs may be essential for the acquisition and defense of nest sites. Pairs stay together outside the breeding season, enabling the extended parental care of juveniles. Paired birds are a powerful social force and may dominate single birds even when acting in the absence of their partner. This helps ensure access to preferred or limited food resources. Among birds generally, genetic studies have shown that most socially monogamous birds engage in some level of extra-pair copulation. Female parrots may be less likely to engage in extra-pair copulations because the costs are high and the benefits small. If a male suspects infidelity he may withdraw parental care, resulting in a failed breeding attempt. The additional genetic input may not be worth this risk given the female will have many more opportunities to reproduce. However, it has been argued that the bright plumage of many socially monogamous species can only be explained by strong sexual selection exerted by birds seeking copulations outside of the pair bond. Determining which of these scenarios best applies to parrots requires further research, with only a few studies completed to date.

Studies of captive Budgerigars *(Melopsittacus undulatus)* have shown that females seek extra-pair copulations when the quality of their male partner is questionable, as measured by his inability to imitate the female's call. Male Glossy Cockatoos *(Calyptorhynchus lathami)* appear to engage in mate guarding, a behavior that limits the access unmated males have to paired females in populations where males outnumber females (male-biased). Genetic studies provide the best evidence for extra-pair paternity (young fathered by other than the breeding male) within populations. Juan F. Masello and colleagues could find no evidence of extra-pair paternity within a colony of Burrowing Parrots *(Cyanoliseus patagonus)* nesting on seas cliffs in Patagonia, Argentina. Anders Gonçalves da Silva and colleagues found no definite cases of extra-pair paternity in native and invasive populations of Monk Parakeets *(Myiopsitta monachus)*. It appears that Burrowing Parrots and Monk Parakeets are sexually monogamous. This is not the case in Green-rumped Parrotlets, where extra-pair males father 8% of all young and 14% of broods contain young fathered by extra-pair males.

Do any parrots have unusual breeding systems?

The Kakapo *(Strigops habroptila)* is the only parrot known to have a "lek" mating system. Among birds generally, a lek is a group of males that gather at an arena to display to visiting females. Each male occupies and defends his own display area, sometimes referred to as a court. Looser aggregations, where males gather in a general area but do not display in sight of each other, are referred to as exploded leks. This is the situation that applies to Kakapo, with courts strung out along ridgelines and extending several kilometers. Arenas are situated to ensure the booming calls of males are heard over a wide area. Each court comprises a number of shallow bowls linked by a network of tracks, all of which are constructed and maintained by the male. Males call from the center of the bowl, which is excavated against a natural sound reflector and windbreak such as a rock or tree. Kakapo males call throughout the night during the breeding season. Their main call is a low frequency booming, achieved by inflating the thorax. Booming may serve to broadcast the location of males and synchronize breeding activity. A higher-pitched metallic call is given between bouts of booming and may assist females in locating specific males. Females visiting the arena mate with one or more males, a small number of males fertilizing the majority of eggs. Males do not attend the nest or assist in the rearing of offspring.

Nest of the Kakapo *(Strigops habroptila)* known as "Alice." It contains her four infertile eggs, plus a three-day-old chick fostered from "Sue" (Stewart Island, March 1985). Don Merton, New Zealand Department of Conservation

Eclectus Parrots *(Eclectus roratus)* exhibit reversed sexual dimorphism, the female being more brightly colored than the male. Rob Heinsohn and Sarah Legge studied the breeding biology of Eclectus Parrots on Cape York, Australia. Females in possession of a nest hollow were attended by up to seven unrelated males, which provided her with food throughout the year and sought mating privileges in return. Only one or two males sire offspring in any given year, but less competitive males may occasionally gain paternity over the course of their lifetimes. Males increase their odds of gaining paternity by providing food to multiple females. The Eclectus Parrot mating system is thought to have evolved due to the male-biased adult sex ratio and the small number of females in possession of nest hollows.

An interesting aspect of Eclectus Parrot reproductive biology is the apparent ability of females to determine the sex of their offspring. Captive Eclectus Parrots are known to produce long unbroken runs of one sex. One captive bird produced 20 males in a row before switching and producing 13 females in a row. Such patterns cannot be explained by chance, and are thought to be the result of females controlling the sex of offspring at the time of fertilization. Rob Heinsohn has shown that wild female Eclectus Parrots may further adjust the sex ratio of their offspring by killing male nestlings soon after hatching (distinguishing the sexes of nestlings is straightforward due to sex-specific down colors). Why do female Eclectus Parrots behave in this way? The circumstances under which brood reduction occurs provide the answer. Females in possession of hollows prone to flooding are more likely to practice infanticide, and only when the brood comprises an older sister and younger brother. Females with poor quality hollows are attended by fewer males, and raising two young can extend the nestling period. The longer the young are in the nest, the greater the risk of nest failure due to flooding. Females fledge several days earlier than males, and it makes sense to favor older sisters over their younger brothers.

Vasa Parrots *(Coracopsis vasa)* exhibit reversed sexual dimorphism, the female being larger than the male and displaying orange skin on her head during the nestling period. Jonathan Ekstrom studied the breeding biology of Vasa Parrots on Madagascar. Their mating system is similar to that of Eclectus Parrots. Male and female Vasa Parrots copulate with multiple partners and nesting females are provisioned by multiple males during the nesting period. Females compete for male parental care, attracting males by singing and chasing females from the nest area. Males have evolved large testes and a unique cloacal morphology that allows long copulations, adaptations that increase the likelihood of eggs being fertilized by their sperm. The factors responsible for the evolution of the Vasa Parrot mating system are unknown. Unlike Eclectus Parrots, a scarcity of nesting hollows does not appear to have played a role. The most likely explanation is food availability, with multiple males required to provide sufficient food for broods.

How long do parrots live?

The average mean maximum lifespan for parrots exceeds 30 years, among the highest for any bird order. Birds may be killed by predators, die of starvation, or succumb to extreme weather events. Death from these causes is referred to as extrinsic mortality. Rates of extrinsic mortality are low in parrots due to their diet, social behavior, and cognitive abilities. Among all birds, herbivores live longer than omnivores or carnivores. Extrinsic mortality rates may be higher in the latter groups due to the direct and indirect risks associated with a predatory or scavenging lifestyle. A high degree of sociality is thought to reduce extrinsic mortality rates because it assists with predator avoidance and improves foraging efficiency. Cognitive abilities improve an individual's capacity to deal with environmental challenges and thus reduce extrinsic mortality rates. Low extrinsic mortality rates mean that parrots may live to an age where adaptations that reduce the effects of aging come under selective pressure. In these circumstances, genes that protect individuals from vascular disease or cancer will be selected. There will be strong selection for longevity as long-lived individuals have more opportunities for reproduction.

There is considerable variation in the maximum recorded lifespan of individual parrots, ranging from 11 years in the Plain Parakeet *(Brotogeris tirica)* to 120 years in the Sulphur-crested Cockatoo *(Cacatua galerita)*. In general, longevity in parrots increases with increasing body size. Larger species are thought to have fewer predators and thus lower rates of extrinsic mortality. Their slower life histories may also be a factor, with delayed maturation postponing the onset of senescence (decline in physiological function associated with ageing). Diet also influences parrot longevity, with granivores living longer than similarly sized species specializing on fruit or nectar. This reflects greater sociality in granivores, but other explanations are possible. Jason Munshi-South and Gerald Wilkinson suggest that periodic food shortage among granivores may extend longevity via caloric restriction and the resultant diversion of resources away from reproduction toward increased body maintenance.

Chapter 7

Foods and Feeding

What do parrots eat?

Seeds dominate the diets of parrots inhabiting temperate and arid environments. These granivorous species come to the ground to feed on grass and herb seed. Many also feed in the canopy, extracting seed from the dry fruits of native shrubs and trees. In tropical environments, parrots include a larger proportion of fruit in their diet. This reflects the greater availability of such foods, many tropical plants enclosing seeds in fleshy rewards to attract vertebrate seed dispersers. While parrots are not usually the intended recipients of this bounty, they make significant inroads into fruit crops. Fruit pulp has been recorded in the diet of many tropical parrots, though it can be difficult to ascertain if birds are feeding on pulp or accessing seeds. Pesquet's Parrots *(Psittrichas fulgidus)* feed almost exclusively on the pulp of figs *(Ficus)*. Birds open the hard outer covering and scoop out the pulp, the fruit remaining attached to the branch throughout. Pesquet's Parrots get around the limited nutritional value of fruit by consuming large quantities, their digestive tracts modified to enable the rapid absorption of nutrients. In addition, they have very low protein requirements. The weak gizzard of Pesquet's Parrots means small seeds pass through the digestive system undamaged, raising the possibility of their dispersing fig seed.

Predominately granivorous or frugivorous parrots include nectar in their diet, especially when it is abundant or other foods are not available. José Ragusa-Netto studied parrot diets in the southern Pantanal, Brazil. Within gallery forest, nectar comprised 30% of the diet of Yellow-chevroned Parakeets *(Brotogeris chiriri)*, rising to 100% in some seasons in some years. On the adjacent savannah, several species of parrot fed on

Pacific Parrotlet *(Forpus coelestis)*.

Heinz Lambert

nectar when the dominant tree, *Tabebuia aurea*, was flowering. The Nanday Parakeet *(Nandayus nenday)* was the most frequent visitor to flowers, while the Blue-fronted Amazon *(Amazona aestiva)* was occasionally sighted. Larger species are probably unable to sustain themselves on nectar alone. Nectar is an important component of the diet of the omnivorous Kaka *(Nestor meridionalis)*, with Ron Moorhouse reporting that birds on Kapiti Island spent up to 16% of their foraging time feeding on nectar or pollen from nine different plants. Kaka also feed on honeydew, a sugary excretion of scale insects. The tip of the Kaka's tongue has a fringe of fine papillae, presumably to aid in the collection of these sugary foods. The tongue of some hanging-parrots *(Loriculus)* is similarly adapted for feeding on nectar.

The lorikeets and the Swift Parrot *(Lathamus discolor)* are nectar specialists and have a number of adaptations to a nectarivorous lifestyle. An elongated tongue with a brush-like tip allows nondestructive harvesting of nectar, while tight glossy plumage prevents feathers from becoming soiled while foraging. Nectar is high in sugar, but low in protein and other essential nutrients. Nectarivorous parrots appear to have lower protein requirements than other species, but may still need to seek additional foods. Pollen is high in protein and a potential food source for birds already visiting flowers in search of nectar. Stephen Hopper and Andrew Burbidge made detailed observations of Purple-crowned Lorikeets *(Glossopsitta porphyrocephala)* feeding on a small-flowered Western Australian mallee eucalypt. Birds positioned the entire flower within the beak and stripped the stamens of pollen using the tongue, possibly collecting nectar at the same time. Active pollen harvesting has been recorded in other nectarivorous parrots and in some parrots with less specialized diets. The Austral Parakeet *(Enicognathus ferrugineus)* relies heavily on pollen from *Nothofagus* flowers during spring and early summer. Many vertebrates have difficulty in extracting nutrients from pollen grains, but Soledad Diaz and Thomas Kitzberger showed that Austral Parakeets achieve pollen emptying rates in excess of lorikeets.

Leaves are an important part of the Kakapo *(Strigops habroptila)* diet. Material may be cut or torn from the plant or chewed in situ. Leaves are crushed against the serrated surface of the upper mandible by the lower mandible. Juice and fine particulates are ingested, with the fibrous waste material left dangling from the plant or ejected from the bill as crescent-shaped wads. Similarly, the leaves of grasses and sedges dominate the diet of Antipodes Parakeets *(Cyanoramphus unicolor)*. Terry Greene reported that grazing by Antipodes Parakeets had resulted in plants being trimmed to the same height over an area of several hectares. Grass leaves comprise a quarter of the diet of Maroon-bellied Parakeets *(Pyrrhura frontalis)* occupying *Araucaria* forests in southeastern Brazil. Other parrots also forage on the nonreproductive parts of plants. Long-billed Corellas *(Cacatua tenuirostris)* and Western Corellas *(C. pastinator)* use their elongated bills to dig up the underground parts of herbs. Pink Cockatoos *(Lophochroa leadbeateri)* pull up annual forbs so they can feed on the succulent tap root. Sap is a significant seasonal component of female Kaka diets, birds chiseling away the bark and then licking sap that leaks from the exposed cambium.

Wood-boring insect larvae are an important food for Yellow-tailed Cockatoos *(Calyptorhynchus funereus)*. Their upper mandible is long and pointed, and the tip of the lower mandible is relatively narrow, characteristics suited for prying insect larvae from their holes and chopping into wood. Exposing the larval gallery can be a difficult task in the absence of a stable work platform. Some populations of Yellow-tailed Cockatoos have developed a technique that allows them to extract wood-boring larvae from smooth-barked trees. Insect larvae are located by the presence of frass holes, small openings through which the larvae eject their waste. A chopping board is constructed by biting into the bark above the frass hole and pulling down and away from the trunk. The bird then perches on this strip of bark while it exposes the larval tunnel and secures its prey.

Kaka also depend on wood-boring invertebrates, with birds on Kapiti Island spending at least 40% of their foraging time procuring this food

“Chews” left by Kakapo *(Strigops habroptila)* feeding on Mountain Flax, Little Barrier Island. Terry Greene, New Zealand Department of Conservation

Adult male Antipodes Parakeet *(Cyanoramphus unicolor)* feeding on *Carex* leaves, Antipodes Island. Terry Greene, New Zealand Department of Conservation

resource. The Yellow-crowned Parakeet *(Cyanoramphus auriceps)* is another New Zealand parrot for which invertebrates are an important source of nutrition. Terry Greene studied Yellow-crowned Parakeets and Red-crowned Parakeets *(C. novaezelandiae)* on Little Barrier Island in the Hauraki Gulf. Invertebrates dominated the diet of Yellow-crowned Parakeets in all seasons, but were of minor importance to Red-crowned Parakeets in most seasons. This dietary separation may have facilitated coexistence of these two closely related species.

Are there any predatory parrots?

The Kea *(Nestor notabilis)* has a reputation for attacking sheep. Perching on the lower back, birds strip the wool and flesh to access fat deposits beneath the skin. While not killed, sheep may succumb later due to infection. Recent video footage has provided evidence of such behavior, though how widespread it is among Kea populations and the number of sheep affected remain open to debate. Conflict between Kea and landholders first emerged when sheep runs were established in the high-country in the latter half of the nineteenth century. At the time, there was an increase in the size of Kea populations due to the ready supply of food in the form of sheep carcasses. Judy Diamond and Alan Bond suggest Kea historically scavenged on moa carcasses, and their innate curiosity reestablished this behavior when large carcasses again became available. The view that Kea caused significant sheep losses led to government sanctioned persecution, with 150,000 individuals killed between 1870 and 1948 under a government bounty scheme. From the middle of last century, improved farm management and the withdrawal of sheep from the high-country brought an end to the conflict. These changes also contributed to the decline in Kea populations, which no longer had easy access to sheep carcasses. Kea are now fully protected, and the major threats are illegal hunting and predation by possums and stoats. Fewer than 5,000 Kea remain, and the species is considered vulnerable to extinction.

The Kea is not the only predatory New Zealand parrot. The Antipodes Islands is a subantarctic group of islands 700 kilometers southeast of New Zealand. They are home to two species of parrot, the Antipodes Parakeet *(Cyanoramphus unicolor)* and Reischek's Parakeet *(C. hochstetteri)*. Terry Greene reported that both species scavenged on the corpses of White-headed Petrels *(Pterodroma lessonii)* and Wandering Albatrosses *(Diomedea exulans)*. He also observed Antipodes Parakeets hunting, killing, and feeding on adult Grey-backed Storm-petrels *(Garrodia nereis)*. Parakeets appeared to systematically investigate burrows within an area. If the burrow was large enough, parakeets would enter and kill the incubating Grey-backed Storm-petrel. Examination of corpses revealed the trachea had been punctured or ripped out, with various muscles and organs eaten. Antipodes Parakeets are significantly larger than Grey-backed Storm-petrels (126–166 grams versus 32 grams).

How much food does a parrot need?

A parrot's diet must include sufficient energy, usually in the form of carbohydrates and fats, to fuel daily activities. Energy is also required for thermoregulation. In addition, the diet must include nutrients necessary for body growth and maintenance. Nutritional requirements will vary with the season, with breeding birds needing additional protein to support egg production and nestling growth. To satisfy dietary requirements, parrots may need to consume a variety of foods in varying quantities. In gathering these foods, birds endeavor to expend as little energy as possible. The costs associated with locating and processing specific types of food are weighed against the benefits derived from its consumption. Parrots will sometimes expend considerable energy accessing foods like insects, presumably because they contain some essential nutrient not readily available elsewhere. The energy expended by Kaka *(Nestor meridionalis)* in excavating wood-boring larvae is thought to exceed the energy gained by consuming the larvae. Such behavior needs to be compensated for by increased consumption of energetically economic food.

Christine Cooper and coauthors examined the metabolic ecology of cockatoos in southwest Western Australia. They calculated the basal metabolic rates of captive birds and used this to estimate the energy requirements of wild cockatoos. These ranged from 534 kilojoules per day for Inland Red-tailed Cockatoos *(Calyptorhynchus banksii samueli)* to 959 kilojoules per day for Butler's Corellas *(Cacatua pastinator derbyi)*. Information on the energy content of seeds from important food plants enabled them to calculate the number of seeds required to meet daily energy requirements. There was little difference in the energy content per gram of the different seeds included in their analyses, meaning the number of seeds that birds had to consume was largely governed by seed size. To meet their daily energy requirements, Forest Red-tailed Cockatoos *(Calyptorhynchus banksii naso)* need to consume the seeds within 753 Jarrah *(Eucalyptus marginata)* fruits or 819 Western Sheoak *(Allocasuarina fraseriana)* cones. However, Red-tailed Cockatoos can process Jarrah fruits twice as quickly as Western Sheoak cones. As a consequence, birds feeding on Jarrah spend half as much time foraging compared to birds feeding on Western Sheoak. Nevertheless, both plants are important foods of Red-tailed Cockatoos. The more energetically costly food may be all that is available at certain times of the year or may contain nutrients not available in more energetically rewarding food. In this regard, sheoaks are noted for their exceptionally high protein content.

Are parrots fussy eaters?

Parrots can be highly selective in their feeding habits. While many environments appear awash with parrot food, investigations of the foraging ecology of parrots often reveal there is less food than first appears. A case in point is the Glossy Cockatoo *(Calyptorhynchus lathami)*, a dietary specialist that feeds on seeds contained within the woody fruits of sheoaks *(Allocasuarina* and *Casuarina)*. Sheoaks are typically abundant in the habitats occupied by Glossy Cockatoos, but are not uniformly exploited. Some trees are heavily fed in, while trees immediately adjacent with abundant fruit are ignored. Intrigued, scientists have investigated tree selection in a number of widely separated populations of Glossy Cockatoos. These studies have shown that birds select feed trees based on the characteristics of their cones, specifically the ratio of seed weight to total seed and cone weight (Clout's Index). Birds feeding on cones with proportionally more seed obtain a greater reward for a given amount of effort. The differences in Clout's Index between feed and nonfeed trees can be small, indicating the existence of a threshold beyond which feeding on sheoak cones ceases to be profitable.

Glossy Cockatoos may be more constrained than most other parrots. Processing sheoak cones is a time-consuming task, raising the possibility that under some circumstances birds may not be able to gather sufficient food during daylight hours to sustain themselves or breed. Tamra Chapman and David Paton did not find evidence for this among Glossy Cockatoos on Kangaroo Island, with nonbreeding birds spending 26% of their time feeding and breeding birds spending 36% of their time feeding. They calculated that nonbreeding males consumed 383 kilojoules per day and breeding males consumed 634 kilojoules per day. It was concluded that abundant food resources meant Glossy Cockatoos spent little time and energy foraging on Kangaroo Island. The situation may be markedly different when food supply is reduced and birds are forced to feed on older cones that are less nutritious and take longer to process. High daytime temperatures over summer may also limit the hours available for foraging. During a period of drought and limited food availability in central New South Wales, I observed a pair of Glossy Cockatoos feeding steadily throughout the day on old cones, resting only briefly at midday. At some point, the quantity or quality of available food will decline to a level that continued occupation of an area becomes untenable.

Other parrots have more flexible diets, allowing them to exploit whatever foods are abundant. Katherine Renton investigated the diet of Lilac-crowned Amazons *(Amazona finschi)* in western Mexico. Birds fed on 33 different species throughout the year, with different foods eaten in different seasons. The number of foods included in the diet expanded and contracted, with a wide variety of foods consumed when food was abundant. Birds tracked the availability of different food resources, switching from one to another depending on availability. The dietary flexibility exhibited by the Lilac-crowned Amazon was thought to be a response to spatial and temporal variability in seed production, which may have evolved to limit the depredations of seed predators such as parrots. Greg Matuzak and colleagues reported similar findings for several parrots in western Costa Rica. Overall, they recorded parrots feeding on 61 types of plant, each parrot species typically feeding on around 30 different types. They found that smaller-bodied species consumed more fruit pulp and flowers than did larger species, a finding they attributed to the greater need of small and large species for energy and protein respectively.

How do parrots find food?

Foraging parrots make decisions at multiple spatial scales. I studied habitat selection by foraging Glossy Cockatoos *(Calyptorhynchus lathami)* in central New South Wales. In the study area, Glossy Cockatoos feed on the cones of two species of shrubby sheoak. Cones are held in the canopy for a number of years. Feeding birds produce a characteristic feeding sign (chewings) that accumulates beneath feed trees. These traits meant it was relatively easy to quantify the available food resource and determine usage by Glossy Cockatoos. I found site use did not depend on the amount of foraging habitat in the vicinity, presumably because birds are able to move between areas without incurring large energy costs. Site selection was based on food supply, birds favoring sites with abundant sheoak and dominated by the more profitable of the two feed species. Having selected a site, birds feed in trees with large cone crops and trees having cones with a high Clout's Index (ratio of seeds to total seed and cone weight). Within a tree, birds preferred to feed on the youngest cones, which were more nutritious and easier to process.

In Botswana's Okavango Delta, Meyer's Parrots *(Poicephalus meyeri)* have been reported feeding on 70 different food items. Seeds are preferred, though other foods are taken and can be seasonally important. The vast majority of feeding takes place high in the canopy. As part of a broader study into the ecology of the species in the region, Steve Boyes investigated how birds located food resources. He found that foraging activity tracked the relative abundance of foods that were visible from the air, highlighting the importance of aerial surveillance in locating food resources. Many of the most significant foods had bicolored displays, increasing their visibility to overflying parrots. Social behavior was thought to play a part in tracking the availability of hidden foods, such as fruits containing insect larvae. In New Guinea, Eleanor Brown and Michael Hopkins investigated how temporal and spatial patterns of flowering in rainforest trees influenced the behavior of nectarivorous birds. They found lorikeets were associated with trees that flowered unpredictably and produced abundant flowers, and sug-

Meyer's Parrots *(Poicephalus meyeri).* Patrick Kelly

gested birds would need to range over large areas to access these scattered resources.

Food abundance and quality are not the only factors influencing foraging decisions. Habitat patches with plenty of food may be avoided if their exploitation poses an unacceptable predation risk. I found that Glossy Cockatoos limited the time spent feeding in open sites, presumably because of the increased risk of predation in these environments. Some parrot food resources are located too far from other critical resources to be utilized. The need to return to the nest each evening or throughout the day limits the area over which breeding birds can forage. In arid environments, parrots need to drink at least once a day and therefore must forage within commuting distance of surface water. In combination, these factors mean the food available to parrots is often less than the total amount of food present in the environment.

Why do some parrots eat soil?

Parrots eat soil to obtain sodium. This finding comes from research undertaken in southeastern Peru, a region containing a relatively high density of clay licks (locally referred to as *collpas*). Soil collected from *collpas* was found to have significantly higher levels of sodium than soil collected from similar sites not visited by parrots. Soils in the Amazon basin are low in sodium, leading to low levels of sodium in plants. As a consequence, the diet of many parrots may be deficient in this essential nutrient. Ingesting soil may be a way of overcoming any shortfall, with *collpa* soils containing 33 times the amount of sodium found in fruit and seeds consumed by parrots.

Ingesting soil may also protect against dietary toxins. Clay particles have a large surface area and a negative charge, allowing them to bind with positively charged toxins and prevent their digestion. *Collpa* soils invariably contain high concentrations of clay, though not in sufficient quantities to explain their selection over other exposed riverbank sites. It appears that parrots select *collpas* based on sodium content, though birds may also benefit from the detoxifying effects of ingested clay. Clay licks are threatened by human disturbance, the clearing of vegetation, and changes to river hydrology. Protection and active management of clay licks is necessary to ensure parrot populations have continued access to this important resource.

Locations where birds can safely access soil are limited, causing them to congregate at a small number of accessible sites. These are typically earth cliffs bordering streams and rivers. Some clay licks have become important tourist destinations as they provide an opportunity for people to observe hundreds of parrots from several species in a single morning. One of the largest and best known clay licks is situated at the Tambopata Research Center (TRC) in southeastern Peru. Here, researchers have observed up to 1,700 parrots from 17 species in a single day. Clay licks of varying size and importance have been recorded elsewhere in the Neotropics. In Bolivia,

Mealy Amazons *(Amazona farinosa)* and Red-bellied Macaws *(Orthopsittaca manilata)* jostle for access to the best clay, Tambopata Research Center, southeastern Peru. Matt Cameron

parrots congregate at cliffs surrounding the Valle de la Luna to consume soil. A daily maximum of 1,044 parrots from five species has been reported for this site. Outside the Neotropics, parrots use clay licks in Australasia and Africa. At Crater Mountain in Papua New Guinea, local landowners have observed several species of parrot feeding on soil. Researchers used remote cameras to investigate these reports, capturing images of Palm Cockatoos *(Probosciger aterrimus)* consuming soil from a steep earth bank within the forest.

Parrot activity at clay licks varies dependent on the weather and season. Weather has a big impact on use of the TRC *collpa*. Fewer parrots visit the lick when it's foggy or raining. This may reflect increased predation risk or reduced foraging activity. Use of the TRC *collpa* is also highly seasonal, with lick use increasing during the breeding season. This reflects seasonal changes in parrot abundance, but also suggests fluctuating requirements for *collpa* soils. Scarlet Macaws *(Ara macao)* feed clay to their young chicks and as a consequence visit the *collpa* more frequently early in the nestling period. Seasonal changes in diet may alter the level of sodium or toxins consumed by birds. In some seasons, the benefits derived from visiting clay licks may be outweighed by the costs. There is much to learn about the pattern of clay lick use in individual birds. Counts at many sites indicate birds do not return on a daily basis. Observations of clay licks within the breeding range of the Maroon-fronted Parrot *(Rhynchopsitta terrisi)* found only a small proportion of the population visited clay licks each day. At the TRC, a reduction in the number of birds visiting the lick due to bad weather is not compensated for by increased numbers of birds using the lick in the following days. This suggests birds are not compelled to visit the lick at regular intervals.

Chapter 8

Parrots and Humans

Do parrots make good pets?

Whether parrots make good pets depends on the establishment of appropriate relationships between humans and birds. Many parrot books, magazines, and websites devote large sections to behavioral problems, highlighting the potential for things to go wrong in parrot-human interactions. Further evidence for this state of affairs is provided by the existence of numerous parrot refuges that take in and seek new homes for birds whose owners can no longer look after them. The cognitive abilities and social nature of parrots contribute to their desirability as companion animals, but these traits make them susceptible to behavioral problems. Looking after a pet parrot requires a considerable investment in time and energy, a commitment that may be required for decades given the longevity of many species. Despite the pitfalls and effort required, many people establish successful relationships with companion birds. These are underpinned by an appreciation for the physical and behavioral needs of parrots and a commitment to parrot welfare.

The types of problems encountered by parrot owners vary. Some owners have simply failed to do adequate research prior to obtaining a bird, and are surprised to find that even well-adjusted parrots can be noisy, destructive, and messy. One of the biggest problems with companion parrots is aggression. Parrots bite people if they feel threatened, especially when they have no means of escape. Birds can become conditioned to bite when such behavior is rewarded by the parrot gaining control of its environment. Parrots may attack people to defend a human with whom they have formed

"Mascots have become part and parcel of a serviceman's life abroad. At Bien Hoa, Vietnam, this parrot recently waddled into the mess hut of RAAF and Army members of No. 161 Independent Reconnaissance Flight and stayed for lunch." Photograph taken by E. Connelly in March 1966. Australian War Memorial, ID number MAL/66/0024/23. The parrot perched on Leading Aircraftman Keith Bell's hand is a Red-breasted Parakeet *(Psittacula alexandri)*.

Australian War Memorial

a close bond. This can disrupt households when everyone but the favored person is attacked, or when a new person moves into a single-person household. Screaming is often cited as a reason for people giving up parrots, while feather plucking can be distressing for parrot owners. Both behaviors may signal boredom or frustration in captive birds. People seem to have more trouble with larger parrots, partly because size amplifies some behavioral problems. Larger species also have an extended period of immaturity, which may increase the likelihood of behavioral problems developing. The more people understand the ecology and behavior of wild parrots, the greater their capacity to meet the needs of captive birds.

People considering purchasing a parrot should become familiar with common behavioral problems and ways they can be prevented. Behavioral problems may take years to emerge and can be difficult to rectify once established. Many parrot behavioral experts caution against allowing pair bonds to develop between parrots and humans. They recommend minimizing activities that encourage the formation of such bonds (e.g., excessive petting and cuddling) and replacing them with appropriate training regimes. Keeping parrots in pairs or groups may assist in reducing problem behaviors. Some parrot owners are reluctant to adopt this approach as they fear it may impact on their relationship with individual birds. Research undertaken by Cheryl Meehan and others has shown this need not be the case, with pair-housing of young parrots in same-sex pairs resulting in improved interactions with human handlers. In any event, it is important that welfare considerations come before the emotional needs of parrot owners. A growing number of owners are allowing companion birds to form within species pairs and breed regularly.

What is the best way to take care of a pet parrot?

It is important that parrots have access to foods necessary to maintain health and vigor. Wild parrots include a variety of foods in their diet, and mirroring this approach in captive birds will help ensure they receive adequate nutrition. Birds are attracted to certain types of foods because they enhance survival in the wild, but allowing captive birds to choose what they eat is unlikely to result in a balanced diet. Introducing new foods to your parrot and maintaining a varied diet can be a time-consuming task but is a necessary part of responsible pet ownership. Apart from its nutritional value, food plays a role in preventing the development of abnormal behaviors. Wild parrots spend long periods of time locating and processing food, while captive birds supplied with seed or pellet mixes spend little time feeding. Feather plucking in parrots may be redirected foraging behavior. Adding fruits, nuts, and vegetables to the diet requires birds to use multiple feeding techniques and to feed for longer periods. This effect can be enhanced by providing food in containers that require problem-solving or additional effort to obtain the food. Cheryl Meehan and coworkers experimentally tested this approach on young Orange-winged Amazons *(Amazona amazonica)*. They found dietary enrichment prevented the development of feather-picking behavior and was effective in reducing the incidence of feather-picking in birds with this problem.

While it is impossible to replicate their natural habitat, every effort should be made to provide interactive environments for captive birds. Some parrots engage in repetitive behaviors that have no obvious function (e.g., bar biting, pacing, manipulation of objects). These can be difficult to detect as they may occur in the owner's absence. Stereotypical behavior can have a number of causes, one of which is lack of physical complexity in the cage environment. Cheryl Meehan and coworkers found that increasing the physical complexity of cages significantly reduced stereotypical behavior in Orange-winged Amazons. Large cages allow parrots greater mobility and can more easily accommodate toys and branches. The regular introduction of novel objects helps ensure that parrots remain stimulated and active. It is possible to purchase nondestructible items and provide these on a rotational basis, though parrots also love to dismantle everyday objects from the home and garden. Care needs to be taken to ensure that provided items are nontoxic and do not pose a risk when taken apart. Researchers at the University of California have shown that color, hardness, size, and material influence the use of cage toys by parrots, and that males and females prefer different types of toys. If your parrot shows no interest in the toys you provide, try something different. Ideally, parrots should have access to a large outdoor flight. This allows birds to stretch their wings and provides greater scope for environmental enhancement. Flights also provide an opportunity for parrots to socialize and engage in typical flock behavior.

What should I do if I find an injured or orphaned parrot?

If you encounter a parrot on the ground and it does not fly away when approached it may be sick or injured. The exceptions are recently fledged parrots, which sometimes end up on the ground when first leaving the nest. Grounded fledglings will usually clamber into adjacent vegetation and continue to be looked after by their parents. They do not require assistance unless they are in immediate physical danger, in which case they can be picked up and placed at a safe height in nearby vegetation. Monitoring the behavior of grounded birds will allow you to determine if they require assistance. Sick or injured parrots can be captured by placing a towel around them, taking care to avoid the powerful bill and sharp claws. Captured birds should be placed on bedding in a dark, well-ventilated box for transport to the nearest wildlife rehabilitation center or veterinary clinic that treats wildlife. Avoid checking the bird en route and do not offer food or water. Unfortunately, if a parrot is sick enough to be captured it usually means it will not respond to treatment. Parrots may carry diseases that can be transmitted to humans. While the risk of infection is low, it's important to practice good hygiene when handling wild birds. Use disposable gloves and avoid touching your face. A disposable face mask provides an additional level of protection. Once you have finished handling the bird, wash your hands thoroughly and launder soiled clothing separately.

How can I see parrots in the wild?

Many people travel long distances to observe parrots in the wild. The selected destination will depend on whether they want to see a particular species or as many different parrots as possible. Timing can be important, as the probability of encountering some species varies with the season. Conservation-minded travelers often travel and stay with tourism operators who can demonstrate their ecotourism credentials. This entails not only the generation of income from nature-based activities, but also protecting habitat and supporting local communities. Ecotourism ventures that host scientific research can use the results to better manage the environment and enhance the experience of visitors. Travelers should satisfy themselves as to the likelihood of seeing parrots and the conditions under which observations will be made. People are sometimes disappointed when they fail to obtain photos similar to those published in wildlife magazines but need to understand these are taken by professional photographers with a lot of experience.

Cities and towns provide opportunities for watching parrots. In Australia, it's common to encounter several species in gardens and parks. These are often abundant, having benefited from man-made changes to the environment. Some lorikeets have gained enormously from the planting of nectar-producing plants in suburban areas, with one study finding that Rainbow Lorikeets *(Trichoglossus haematodus)* were present in 76% of gardens in the Greater Sydney Region. Justine Smith and Alan Lill studied the foraging ecology of Rainbow Lorikeets and Musk Lorikeets *(Glossopsitta concinna)* in Melbourne, Victoria. Lorikeets mostly fed on native plants not indigenous to the Melbourne area, with a few species of ornamental eucalypt particularly important. These trees had been planted for the duration and consistency of their flowering, an aesthetic judgment that inadvertently benefited lorikeets. Ground-feeding granivorous parrots have also benefited from urbanization, the establishment of lawns, parks, and sports fields providing ample foraging habitat. A study of parkland sites in Melbourne found that common turf grasses provided more than half the diet of Red-rumped Parrots *(Psephotus haematonotus)*. Mirroring the results of other studies, Red-rumped Parrots did not appear to compete for food, suggesting it was abundant. A shortage of nest sites is more likely to limit parrot populations in urban areas, birds relying on bushland fragments and remnant trees.

Some parrot species have established populations in cities outside regions where Psittaciformes naturally occur, providing incongruous opportunities for wildlife watching. Of these, the Rose-ringed Parakeet *(Psittacula krameri)* has been the most successful, with populations reported in 35 countries on four continents. They are widespread throughout Europe, and can be readily observed in Greater London. The Monk Parakeet *(Myiopsitta monachus)* has been similarly successful in establishing itself from deliberate or accidental releases outside its natural range. Populations are dotted throughout Europe and the United States, with an accessible and much publicized population in New York. In the United States, communities of mostly Neotropical parrots exist in a number of southern states, possibly because the climate in these areas more closely matches their native range. In southern California, 13 species of parrot have established populations, with flocks commonly encountered in the Greater Los Angeles area. In Florida, 17 species of parrot have been reported as breeding, though not

Table 8.1. Parrot species known to have established populations in California

Common name	Scientific name
Rose-ringed Parakeet	*Psittacula krameri*
Blue-crowned Parakeet	*Aratinga acuticaudata*
Mitred Parakeet	*Aratinga mitrata*
Red-masked Parakeet	*Aratinga erythrogenys*
Nanday Parakeet	*Nandayus nenday*
White-winged Parakeet	*Brotogeris versicolurus*
Yellow-chevroned Parakeet	*Brotogeris chiriri*
White-fronted Amazon	*Amazona albifrons*
Red-crowned Amazon	*Amazona viridigenalis*
Lilac-crowned Amazon	*Amazona finschi*
Red-lored Amazon	*Amazona autumnalis*
Blue-fronted Amazon	*Amazona aestiva*
Yellow-headed Amazon	*Amazona oratrix*

Source: www.californiaparrotproject.org/id_guide.html, accessed 4 June 2011.

all of these species have populations large enough to be considered self-sustaining.

A good pair of binoculars is an essential piece of equipment for watching parrots, irrespective of the environment or conditions. A spotting scope (telescope) allows closeup views of birds hidden in the canopy or resting on a distant cliff. Obtaining good parrot photographs usually requires a quality DSLR camera and telephoto lens (i.e., at least 300 millimeters). A tripod helps ensure images are sharply in focus, especially when there is limited light—a frequent occurrence. An adaptor can be used to attach a digital camera to a spotting scope, allowing photographs to be taken through the spotting scope eyepiece (digiscoping). Many environments in which parrots occur are tough on camera equipment, with the high humidity of tropical environments a particular concern. Storing equipment overnight in sealable plastic bags with silica gel helps prevent damage caused by excess moisture. Favorite images can be backed up onto spare memory cards or some other form of digital storage (e.g., portable hard drive or laptop). Electrical adaptor plugs, spare battery charger, and spare batteries should be carried, with every opportunity taken to recharge the latter. It is important to become familiar with camera equipment before traveling.

Should people feed wild parrots?

There is debate as to whether backyard feeding is benign, beneficial, or detrimental to wild bird populations. Opponents of backyard feeding express concern about the potential for birds to become reliant on handouts, reducing their natural fitness and making them vulnerable if artificial foods are withdrawn. They note that common species benefit from backyard feeding, possibly at the expense of rarer species. Feeding is also said to facilitate the spread of disease, birds gathering in large numbers at what are often unhygienic feeding stations. Proponents of backyard feeding highlight the opportunity it provides for people to reconnect with nature and develop positive attitudes toward wildlife and conservation. They point to the role backyard feeding plays in helping birds through the winter or other periods of food shortage.

There is little information on the costs and benefits to wild bird populations of backyard feeding. Most feeding of native parrot populations occurs in Australia, but goes unnoticed except when birds make a nuisance of themselves. Populations of common parrots with access to feeders presumably benefit from the extra food, perhaps breeding more frequently or producing more fledglings per nesting attempt. Research is needed to confirm this and to ascertain the level of impact on species that avoid feed stations. A number of disease outbreaks in local populations of Australian parrots have been linked to the use of feeding stations, and these may cause a reduction in parrot numbers.

In the Northern Hemisphere, feeding of introduced parrots occurs and has been controversial in some cities. The persistence of some populations is dependent on access to backyard bird feeders, with one study finding that Monk Parakeets *(Myiopsitta monachus)* in Chicago fed exclusively on

There is a long tradition of tourists feeding Rainbow Lorikeets *(Trichoglossus haematodus)* at Currumbin Wildlife Sanctuary, southeast Queensland. Matt Cameron

birdseed over the winter months. The City and County of San Francisco banned the feeding of Red-masked Parakeets *(Aratinga erythrogenys)* and other wild birds in public places in 2007, a decision born out of concern for the welfare of the local parrot flock. There were a number of reasons for this concern, including the potential for people to be injured or become ill and the risk of birds being captured for the pet trade.

For people who choose to feed birds in their backyard, Michelle Plant has developed a simple best practice guideline that aims to minimize negative impacts on wild birds. This guideline is relevant to the feeding of wild parrots. Just like captive birds, wild parrots require a balanced diet. Select quality seed mixes, and avoid feeding large quantities of high-fat seeds (e.g., sunflower). Supplement seed with fruit, vegetables, and nectar. Feeding stations should be designed to minimize fecal contamination and the amount of food falling onto the ground. Feeding dishes should be removable to allow regular cleaning and sterilization. Provide small quantities of food once or twice a day and remove food at the end of each feeding session. Putting food out on alternate days will help prevent parrots becoming reliant on artificial food. Skipping a few days or weeks should not be a problem for birds with predominately natural diets. Site your feeding station so parrots can visit it safely and neighbors will not be disturbed.

Parrots grace the stamps of many countries. This set of five stamps featuring Australian parrots was issued in 2005.
Reproduced with permission of Australia Post

Keep the area clean and clear of spilt food to avoid disease and pest issues. To minimize the risk of contracting diseases carried by parrots, wash your hands thoroughly and never bring feeding equipment into your kitchen or parts of the house where food is consumed. An alternative to feeding stations is to develop a parrot-friendly garden that provides the seeds, fruits, insects, and nectar required for a balanced diet. This may prove to be a more sustainable way of ensuring the long-term health of the local parrot population.

Peregrine Falcon *(Falco peregrinus)* feeding on Galah *(Eolophus roseicapillus)*. Lindsay Cupper

Kea *(Nestor notabilis)*. Matt Cameron

Red Lory *(Eos bornea).* Heinz Lambert

Grey Parrot *(Psittacus erithacus).*
Heinz Lambert

Red-bellied Parrot *(Poicephalus rufiventris)*. Heinz Lambert

Meyer's Parrots *(Poicephalus meyeri)*. Patrick Kelly

Juvenile Cape Parrot *(Poicephalus robustus)* feeding on pecan nut.

Heinz Lambert

Black-winged Lovebird *(Agapornis taranta)* feeding nestling.

Heinz Lambert

Lilian's Lovebirds *(Agapornis lilianae).* Phil Perry

White-winged Parakeets *(Brotogeris veriscolurus).* Heinz Lambert

Scarlet Macaw *(Ara macao).* Roy Toft, www.toftphoto.com

Hyacinth Macaws *(Anodorhynchus hyacinthinus).* Roy Toft, www.toftphoto .com

White-bellied Parrot *(Pionites leucogaster).* Heinz Lambert

Hyacinth Macaw *(Anodorhynchus hyacinthinus)*, Piauí, Brazil.

Matt Cameron

Thick-billed Parrots *(Rhynchopsitta pachyrhyncha)*, western Chihuahua, Mexico. Steve Milpacher

Orange-cheeked Parrots *(Pyrilia barrabandi)* and Blue-headed Parrots *(Pionus menstruus)* at clay lick. Heinz Lambert

Red-and-green Macaws *(Ara chloropterus)* at clay lick. Heinz Lambert

Chapter 9

Parrot Problems (from a human viewpoint)

Are some parrots agricultural pests?

I pulled up next to the fruit packing shed and went in search of the landholder. I had been asked to assess the damage done to his apple crop by Gang-gang Cockatoos *(Callocephalon fimbriatum)*, a threatened parrot endemic to southeastern Australia. Following a brief discussion about the weather, the farmer took me to look at the block where the birds had been feeding. A few rows along the farm boundary had been badly impacted, the ground littered with bits of apple removed by the birds in their quest for seeds. Talking to the landholder, it became apparent he had made no effort to harass the birds when they began visiting his orchard. The initial damage was slight and he was generally sympathetic to wildlife. Once the level of damage had become a concern, ad hoc efforts to scare the birds off the crop had failed.

I hung around to await the arrival of the cockatoos late in the afternoon. Right on time, 15 birds flew into a stand of remnant vegetation bordering the orchard. They paused briefly before dropping down into the apples to feed. Without allowing them time to settle, I chased them out of the orchard. The startled birds sheltered in nearby eucalypts, but soon returned to the crop. This time, I allowed them to feed for 30 minutes before attempting to scare them. Moving the birds out of the orchard proved much more difficult on this second attempt. They were more inclined to shift to a different part of the orchard and continue feeding than retreat to the safety of the remnant vegetation. The message from this simple experiment seemed clear. Allowing birds to settle on the crop and feed for any length of time reduced the effectiveness of harassment techniques and could result in increased crop damage.

I wandered back up to the farmhouse and sought out the landholder. We sat down and discussed the various options for minimizing crop damage. I explained that now the cockatoos had developed the habit of feeding in his orchard it would be difficult to alter their behavior. Commonly employed scaring techniques (e.g., explosive cartridges, gas guns, and harassment by vehicle) would likely work, but needed to be part of a well-thought-out pest management strategy. He needed to be committed to maintaining the scaring program until the threat posed by the cockatoos no longer existed. Netting was an option and had the advantage of being effective in the absence of people. Many of the surrounding orchards had been netted to prevent hail damage, which had the additional advantage of reducing the impact of bird pests. Ultimately, the methods to be employed would depend on the cost of control versus the cost of lost production. In this situation, the level of damage caused by the cockatoos appeared insufficient to warrant control.

Within Australia, a wide variety of crops are affected by parrots. The cost of parrot damage varies between farms and years. The impact at a regional level is typically small. Parrots are a serious pest of horticultural crops, consuming buds, flowers, and fruit. In South Australia, Crimson Rosellas *(Platycercus elegans)* foraging in cherry orchards can remove 10–90% of the buds from trees within a block. Cockatoos are a significant pest of grain and oilseed crops, birds feeding on germinating seed, standing crops, and stubbles. The propensity of cockatoos to form large flocks means they are preadapted to the exploitation of abundant, ephemeral food resources. Different crop stages may be exploited by different cockatoo species, while some crops are favored over others (e.g., oilseeds are preferred to grains). In addition to their direct impact on yields, parrots can damage agricultural infrastructure. Cockatoos attracted to peanut farms in northern Queensland make use of their powerful bill to chew through the electrical wiring of pivot irrigators. This results in irregular irrigation and subsequent yield reductions. In addition, there is a cost associated with the repair or modification of equipment to make it less vulnerable to cockatoo damage (e.g., steel casings for wiring).

In the Neotropics, parrots have long been considered agricultural pests. Enrique Bucher noted that pre-European cultures in South America had problems with parrots attacking corn plantations. Charles Darwin commented on the status of Monk Parakeets *(Myiopsitta monachus)* as agricultural pests: "These parrots always live in flocks, and commit great ravages on the corn-fields. I was told that near Colonia 2,500 were killed in the course of one year." Neotropical parrots exploit a similar range of crops to

Adult male Gang-gang Cockatoo *(Callocephalon fimbriatum)* eating apple seeds, Batlow, New South Wales. Matt Cameron

Table 9.1. Australian parrots reported as causing serious damage to horticultural crops

Common name	Scientific name
Baudin's Cockatoo	*Calyptorhynchus baudinii*
Yellow-tailed Cockatoo	*Calyptorhynchus funereus*
Galah	*Eolophus roseicapillus*
Sulphur-crested Cockatoo	*Cacatua galerita*
Little Corella	*Cacatua sanguinea*
Western Corella	*Cacatua pastinator*
Long-billed Corella	*Cacatua tenuirostris*
Rainbow Lorikeet	*Trichoglossus haematodus*
Scaly-breasted Lorikeet	*Trichoglossus chlorolepidotus*
Musk Lorikeet	*Glossopsitta concinna*
Australian Ringneck	*Barnardius zonarius*
Crimson Rosella	*Platycercus elegans*
Eastern Rosella	*Platycercus eximius*
Western Rosella	*Platycercus icterotis*

Source: Tracey et al. (2007).

Australian parrots, but also compete with people for extractive bush foods. Mandar Trevidi and coauthors investigated the impact of large macaws on a concession crop of Brazil Nuts *(Bertholletia excelsa)* in southeastern Peru. Macaws destroyed around 10% of the crop, a loss additional to that caused by agouti *(Dasyprocta)* predation and spoilage. These losses were swamped by price fluctuations, but ecotourism and certification of the crop as "wildlife friendly" were suggested as possible methods for compensating concession holders. The provision of corn has been trialed as a mechanism for compensating subsistence farmers for damage caused by Lear's Macaws

(Anodorhynchus leari). It was felt that compensation would reduce the likelihood of landholders taking retaliatory action against this endangered parrot. Unless strategies are put in place to minimize damage to crops, this type of compensation needs to be maintained in perpetuity if long-term negative outcomes are to be avoided.

Craig Symes grew up on a farm near Creighton in the KwaZulu-Natal midlands of South Africa, a region that historically supported large numbers of the now critically endangered Cape Parrot *(Poicephalus robustus).* In an article for Czech parrot magazine *Papoušci,* Symes reflected on the relationship between his family and Cape Parrots. His grandfather, who began farming in the region in the 1920s, could recall seeing flocks containing hundreds of birds. These flocks visited local orchards, including those attached to Centocow, a Catholic mission. Centocow had been built in the shadow of Hlabeni Mountain, whose mist-shrouded *Podocarpus* forests provide foraging and nesting habitat for Cape Parrots. The Trappist monks who labored in the mission's orchards considered the parrots a pest. Symes told the story of his father, then a young boy, being rewarded by the monks for each parrot he shot. The scale of these control efforts is unknown, though Symes' father once observed a pile of at least 100 dead parrots. In the early 1990s, Olaf Wirminghaus commenced a study on Cape Parrots in the area around Creighton. Symes, who was caring for an injured Cape Parrot at the time, came to the attention of Wirminghaus and was employed on the project as a research assistant. His knowledge of the parrot's local haunts, including Hlabeni Forest, proved invaluable. Following completion of the Cape Parrot project, Symes went on to study the Greyheaded Parrot *(P. fuscicollis suahelicus).* He continues to work for the conservation of African parrots.

Accurately quantifying the damage caused by parrots is problematic. Farmer surveys are often used, but these overestimate the impact parrots have on crop yield. This is because parrots are the most visible pest species. Most studies on parrot crop damage report a high degree of variability in impacts experienced within and between farms. Some blocks or farms are badly impacted, while others suffer little damage. Sites in close proximity to roosts or refuge areas are more likely to be affected. Birds are more likely to settle at sites receiving limited human visitation. Variability in the level of damage between years can often be related to changes in the abundance of natural foods. In southwest Western Australia, damage to horticultural crops caused by a number of parrots is greatest in years when Marri *(Eucalyptus calophylla)* fails to flower or set seed. Monitoring the availability of natural foods may assist in predicting years when there is the potential for parrots to cause significant damage. This will allow appropriate measures to be put in place to minimize crop losses.

Can parrots be a nuisance in urban areas?

Their powerful bills and gregarious nature mean cockatoos can make a nuisance of themselves in urban areas. Soft-timbered houses or outdoor furniture are especially vulnerable. In one well-publicized case, a small group of Sulphur-crested Cockatoos *(Cacatua galerita)* did extensive damage to a house built of decorative timber while the owner was away for two days. Several other houses were also damaged. It turned out the birds had been attracted to the area by grain stored in a neighbor's shed. In these situations, removing the free food will often solve the problem. Parrots are capable of damaging urban infrastructure. Within and outside its natural range, the Monk Parakeet *(Myiopsitta monachus)* builds stick nests on electrical utility poles. These provide a sound platform for nest construction, but birds may also be attracted by the heat given off by electrical equipment. Nests can prevent adequate ventilation of equipment, causing it to short-circuit and catch fire. The impact on Monk Parakeets of nest removal can be minimized by providing alternate nest platforms and ensuring nest removal is undertaken at a time when birds are neither breeding nor reliant on the nest for shelter.

Are parrots ever invasive species?

Many parrots have established themselves outside their natural range, largely as a result of deliberate or accidental releases of captive birds. Unassisted colonization of some areas may have occurred, with humans playing a role by creating favorable habitat conditions. Two parrot invaders have received widespread attention—the Monk Parakeet *(Myiopsitta monachus)* and the Rose-ringed Parakeet *(Psittacula krameri)*. The Monk Parakeet is a native of southeastern South America. It has established breeding populations in the United States, Europe, Japan, Kenya, and Puerto Rico. A recent genetic study suggests North American populations are derived from the subspecies found in southeast Brazil, Uruguay, and northeast Argentina *(M. m. monachus)*. This region supplied most of the birds for the international pet trade, confirming the industry as the most likely source of naturalized populations. Rose-ringed Parakeets are naturally distributed through central Africa and southern Asia. They have established populations in North America, Europe, the Middle East, East Africa, southern Africa, Southeast Asia, and Japan. Releases and escapes are thought to be responsible for the establishment of most of these, though the origin of many is uncertain and some are potentially natural.

Diederik Strubbe and Erik Matthysen have investigated factors leading to the successful establishment of Monk Parakeets and Rose-ringed

Parakeets in Europe. They gathered evidence of 167 separate Rose-ringed Parakeet introductions and 90 separate Monk Parakeet introductions, all from areas of human settlement. They were able to confirm that males and females were released in 123 and 58 cases respectively. Around half of documented introductions were successful. The likelihood of parakeets establishing themselves was greater in areas with high human population densities. Within their natural range, these parakeets occupy a variety of habitats and can occur in agricultural and urban environments. This ecological flexibility has allowed them to exploit urban environments in Europe, which provide abundant food and suitable nest sites. The potential for introductions to be supplemented by further releases is greater in more densely populated regions. Supplementation reduces the likelihood of small populations going extinct. Most introductions failed in areas having more than 50 frost days per year, possibly due to lowered reproductive success rather than increased mortality from starvation or exposure. Stick nests did not confer any advantage on Monk Parakeets in terms of cold tolerance, though Monk Parakeets appeared better able to cope with the combined effects of cold and rain than Rose-ringed Parakeets.

While some introduced populations of Rose-ringed Parakeets and Monk Parakeets appear stable, many are increasing in size and spreading. Monk Parakeet populations in the United States have demonstrated their capacity for exponential growth, with the number of birds recorded during Christmas Bird Counts in Florida rising from 35 in 1980 to 3,041 in 2002 (alternatively, 0.0066 to 0.5536 birds/hour). Similarly, the number of Rose-ringed Parakeets using a London roost increased from 2,700 birds in 2001 to 7,000 birds six years later. These increases may reflect low levels of mortality or higher reproductive rates relative to natural populations, though little data are available. A group of scientists from several countries investigated these issues in Rose-ringed Parakeets by studying a native population in India and introduced populations in Israel and the United Kingdom. Introduced populations experienced lower levels of nest predation due to a reduction in the number of nest predators. Egg fertility was low in the United Kingdom relative to the other populations, attributable to colder temperatures. Reduced nest predation may facilitate the growth of introduced Rose-ringed Parakeet populations, though this is mediated in cold climates by low levels of egg fertility.

Monk Parakeets and Rose-ringed Parakeets are considered significant agricultural pests within their native ranges. Concerns about the potential for Monk Parakeets to negatively impact on the United States agricultural sector led to the species being banned in a number of states and early failed eradication efforts. Despite the steady growth in Monk Parakeet popula-

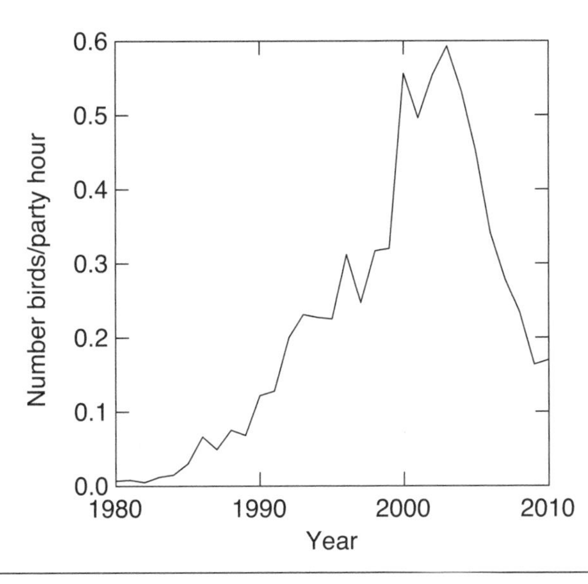

Number of Monk Parakeets recorded during Florida Christmas Bird Counts, 1980–2010. The reason for the recent decline in numbers is not known. Results are presented as the average number of birds recorded in an hour by each party of observers. Data from National Audubon Society Christmas Bird Count Historical Results, retrieved November 29, 2011, from www.audubon.org/bird/cbc.

tions, the potential threat to agriculture in the United States has not yet materialized. Instead, the bird has become a significant issue for electrical utilities who must contend with fires and power outages caused by birds building their nests on electricity distribution infrastructure. The increase in Rose-ringed Parakeet numbers in the United Kingdom is starting to bring parakeets and farmers into conflict, with anecdotal reports of horticulturalists losing up to 20% of their crop to the birds. Invasive species are often linked to the decline of native fauna and flora. Parrots have the potential to compete with other cavity users for a limited hollow resource. Experiments undertaken in Belgium have demonstrated that when cavities are scarce, introduced Rose-ringed Parakeets will displace native Wood Nuthatches *(Sitta europaea)* from their hollows, causing a decline in nuthatch populations.

Electrical utilities have responded to problems caused by Monk Parakeets in a number of ways. The most common technique is the removal of nests, though this requires an ongoing commitment in resources as birds will often commence rebuilding nests in the same location almost immediately. The other technique employed is trapping and removal, with birds usually caught in the nest at night and euthanized. These actions are not always supported by the public, many people considering introduced parrots to be a positive addition to the environment. Accordingly, nonlethal control methods are being investigated. The chemical diazacon (20,25-diazacholesterol dihydrochloride) has been shown to reduce fertility in captive Monk Parakeets and Rose-ringed Parakeets. It works by lowering cholesterol levels and thus inhibiting the production of cholesterol-dependent reproductive hormones. Michael Avery and colleagues examined the impact of diazacon on a free-flying Monk Parakeet population by feeding treated and untreated grain to birds at different sites. They reported a 68% reduction in nest productivity at treated sites and concluded that diazacon had promise as a means of controlling the growth of Monk Parakeet populations. They highlighted the importance of ensuring that nontarget granivores do not have access to feed stations.

Current efforts to control Monk Parakeet populations in the United States are aimed at reducing costs incurred by utility companies and minimizing disruptions to their customers. Eradication of Monk Parakeets is not an objective, with a number of authors considering this impractical given their current distribution and public opposition to lethal control. Other countries take a more precautionary approach to the control of introduced parrots, acting to eradicate populations that pose a risk to agriculture or biodiversity. Like many island biotas, New Zealand's vertebrate fauna has been devastated by invasive species. The government reacted to deliberate releases of Rainbow Lorikeets *(Trichoglossus haematodus)* into the Auckland area by declaring the species an unwanted organism under the country's biosecurity legislation. The population is managed by trapping to very low densities, wildlife authorities accepting that eradication is unlikely given the potential for further releases. Rainbow Lorikeets have also established themselves in metropolitan Perth, Western Australia. The population in 1995 was estimated at around 2,000 birds and is thought to be increasing exponentially. Eradication is not considered feasible, and management is targeted toward reducing the size of the population and preventing its spread beyond the metropolitan area.

Do parrots have diseases and are they contagious?

Psittacosis is a human disease closely associated with the keeping of parrots. It is caused by the bacterium *Chlamydophila psittaci*. The same bacterium causes avian chlamydiosis, a disease known from a large number of bird species. Humans typically become infected via contact with pets and poultry, inhaling the bacterium present in the droppings or secretions of infected birds. Symptoms in birds include diarrhea, discharge from the eyes or nostrils, lack of appetite, and weakness. Humans experience flu-like symptoms, including fever, chills, headache, muscle ache, chest pain, and breathlessness. A study of apparently healthy Amazon parrots from breeding facilities in Brazil found that nearly 36% of birds were shedding *C. psittaci*. A higher percentage of birds had antibodies to the bacterium, suggesting prior exposure. These results are mirrored in other studies. Around 50

Introduced Monk Parakeets *(Myiopsitta monachus)* often build their nests on power infrastructure, sometimes causing fires and blackouts.

Michael L. Avery

confirmed cases of Psittacosis are reported in the United States each year, though diagnosis is difficult and many cases probably go undetected. Psittacosis is treatable with antibiotics, though fatalities do occur. The risk of contracting the disease can be reduced by practicing good personal hygiene and not allowing parrot fecal material to accumulate and become airborne. Mouth-to-beak contact with parrots should be avoided.

Psittacosis and the role played by parrots in its transmission were recognized in the late 1800s, but the disease became notorious during a worldwide outbreak in 1929–1930. This outbreak was thought to have originated in Argentina, with the international transport of infected parrots resulting in the disease spreading to all parts of the globe. Around 750–800 people were infected, with a mortality rate of around 15%. It has been suggested that outbreaks in different parts of the world may not have been connected, but simply reflected heightened awareness of the disease and increased monitoring of captive birds. Psittacosis remains of concern to health authorities, with outbreaks occasionally reported among high-risk groups such as zookeepers and customs officers. Psittacosis outbreaks have also been linked to wild birds. A probable Psittacosis outbreak occurred in the Blue Mountains, an area of extensive bushland immediately west of Sydney. Fifty-nine people were thought to have been infected, exposed to the bacterium when they handled sick birds or mowed their lawn. Researchers investigating this outbreak found the use of a grass catcher when mowing reduced the likelihood of contracting the disease. They recommended the use of personal protective equipment for people working in areas where contact with wild birds and their droppings is likely.

Chapter 10

Human Problems (from a parrot's viewpoint)

Are any parrots endangered?

We had been climbing steadily for an hour, zigzagging across the steep slope. Our objective was the top of a cliff on which nine pairs of Red-and-green Macaws *(Ara chloropterus)* nested. The landholder had arranged for his brother to act as my guide, and a couple of children had come along for the ride. My tortoise-like advance contrasted starkly with the hare-like progress of the children. They nimbly negotiated the worst parts of the climb, rubber thongs apparently no disadvantage. On reaching the summit, I had to clutch at the sparse vegetation and steady myself. I was standing on a small area of level ground atop a vertiginous cliff. I made a slow turn and marveled at the unbroken expanse of vegetation in all directions. I knew people eked out a living in this country, but from my present vantage point it appeared an untouched wilderness. The scene was markedly different from what I had witnessed only a few hundred kilometers south. Flying into Barrieras two days earlier, my aircraft window had framed a view of orderly fields and pivot irrigators. Much of the vegetation surrounding this agricultural center in Brazil's Bahia State had been cleared for soybean production. My thoughts were interrupted by a tug on my sleeve—the children had located a pair of macaws on the cliff. The birds were perched in one of the potholes used for nesting, though it was too early in the season for breeding to be underway. My guide explained they had once climbed down to these nests to retrieve the young; now they preferred to show the birds to tourists. As the landowner later explained, he had been "absent from nature, but now he was present in nature."

Parrots are one of the most threatened groups of birds in the world, with

27% of species listed as threatened. An additional 12% of species are of conservation concern. Parrots have a number of attributes that make them especially vulnerable to extinction. First, many parrots are large-bodied and slow-breeding. In these species, adults live for a long time but produce relatively few young. This sometimes delicate balance between mortality and fecundity can be easily disrupted by small alterations to nest success and survival rates. Second, numerous parrots have small populations or restricted ranges. Populations of large-bodied species may be small because a substantial area of habitat is required to sustain an individual, while parrots of all sizes can have small populations when they are restricted to small areas (e.g., islands). Small populations are more likely to be wiped out by catastrophic events than are large populations. Third, the slow life history of some parrots means they are ill-equipped to recover from any reduction in population size. The longer a population remains small, the greater its risk of extinction. Finally, nearly all parrots rely on cavities for nesting, and some have specialized diets. Specialization on a resource makes a species vulnerable to changes in the abundance or distribution of that resource.

Loss of habitat continues to pose a risk to the world's parrots. Around half of the world's tropical forests have been cleared, and much of what remains is degraded or fragmented. Some of the most important countries for parrot conservation have the worst recent land-clearing records (e.g., Australia, Brazil, Indonesia, Mexico) or are experiencing increased rates of deforestation (e.g., New Guinea, Solomon Islands). In southeastern Brazil, >90% of the Atlantic rainforest has been cleared, with remaining areas existing as isolated fragments. Several parrots are endemic to this area, including the endangered Red-browed Amazon *(Amazona rhodocorytha)* and vulnerable Blue-throated Parakeet *(Pyrrhura cruentata).* In Australia, more than 50% of the vegetation outside the arid zone has been cleared for agriculture, with >500,000 hectares cleared each year between 1990 and 2000. This clearing has been a major factor in the decline of species such as Carnaby's Cockatoo *(Calyptorhynchus latirostris)* and the Swift Parrot *(Lathamus discolor).* A recent study tracking changes in forest cover in Papua New Guinea found that 15% of tropical forests had been cleared between 1972 and 2002, with deforestation rates of between 1–18% recorded for individual bioregions. Much of this deforestation occurred in lowland forests, though high clearing rates were also recorded in forests adjoining subalpine grasslands. Worryingly, these forest types are poorly represented in the relatively small protected area network (2.8% of PNG land area).

Large areas of tropical forest have been commercially logged or are subject to logging concessions. Disturbance can sometimes increase the abundance of parrot food plants, providing a short-term benefit to species able to exploit secondary habitats. However, the decline in nest cavities

Table 10.1. Number of parrot species in each IUCN Red List category

Red List category	Number of parrot species
Critically Endangered	15
Endangered	33
Vulnerable	48
Near Threatened	41
Least Concern	218

Source: IUCN Red List version 2011.1.

Critically Endangered = extremely high risk of extinction in the immediate future. *Endangered* = very high risk of extinction in the near future. *Vulnerable* = a high risk of extinction in the medium-term future. *Near Threatened* = close to qualifying for Vulnerable. *Least Concern* = do not presently meet the criteria for listing as Threatened.

Paddock trees provide nest sites for Red-tailed Cockatoos *(Calyptorhynchus banksii graptogyne)* in southwest Victoria.

Matt Cameron

that is an inevitable consequence of logging means most parrot populations are likely to decline in the longer term. Sustainable forestry operations provide an opportunity for countries and communities to derive an income from native forests, which may provide an incentive for the retention of parrot habitat. The persistence of parrots in well-managed forests, even at lower densities, is preferable to these areas being cleared for agriculture. However, there are rarely sufficient data to allow the development of sustainable forestry practices. Even where these exist, limited regulation and inadequate monitoring make their implementation problematic.

As well as altering the structure and composition of forests, logging contributes to deforestation. One way this occurs is via the construction of logging roads, there being a close association between frontier roads and forest loss. It has been reported that 95% of deforestation in the Brazilian Amazon occurs within 50 kilometers of highways or roads. In a recent review article, Bill Laurance and coauthors suggest the impacts of logging roads can be reduced by concentrating forestry operations in areas served by existing roads, avoiding excessive road construction and closing roads once forestry operations have finished. Increased human activity along roads also increases fire frequency, with logged forests tending to be drier and more easily burnt. Global warming is expected to result in an overall drying of tropical forests and more intense or frequent droughts, factors that will further increase the susceptibility of tropical forests to fire. Deforestation, fire, and global warming are part of a positive feedback loop that will lead to widespread forest loss in the coming decades.

Invasive species are a significant problem for parrots occupying islands, many of which lacked native terrestrial mammals at the time they were colonized by humans. As a consequence, native birds were vulnerable to the environmental changes and increased predation pressure caused by humans and the species they brought with them (e.g., pigs, dogs, rats). David Steadman estimates that 18 species of parrot were lost from the Pacific (excluding New Zealand) following the spread of humans through the region. Rats, in particular the Black Rat *(Rattus rattus)*, threaten the survival of a number of extant island parrots, including the Blue Lorikeet *(Vini peruviana)* and Ultramarine Lorikeet *(V. ultramarine)*. New Zealand's parrots have been particularly hard hit by introduced mammalian predators and competitors. Mustelids, principally Stoats *(Mustela erminea)*, and Common Brushtailed Possums *(Trichosurus vulpecula)* are responsible for declines in populations of Kaka *(Nestor meridionalis)*. Both species eat eggs and nestlings and will kill adult females at the nest. The preparedness of females to enter nests to attack Stoats that have killed their nestlings makes them especially vulnerable. Predation of adult females by introduced predators is thought responsible for the strong male sex bias (minimum three males

Adult female North Island Kaka *(Nestor meridionalis)* with nestlings.
Terry Greene, New Zealand Department of Conservation

to one female) observed in Kaka populations. Fledgling Kaka are poor flyers and spend considerable periods of time on the ground where they are vulnerable to predators. In the absence of pest control, Terry Greene found that 62% of Kaka fledged in one season were killed within two to three days of leaving the nest.

Have any parrot species gone extinct?

Nineteen species of parrot are listed by the IUCN as having gone extinct in recent times (i.e., since 1500). This is an underestimate, as many species disappeared before attracting the attention of early explorers or travelers. There is debate whether some extinct species actually existed, as not all of them are known from museum specimens. Paintings exist for some species, others are known from early writings. Drawings and descriptions were not necessarily based on first-hand observations, leaving scope for confusion as to the existence and appearance of species. Doubt about the existence of some birds has led to the term "hypothetical species" being coined. Many extinct parrots occurred on islands, continuing a prehistoric trend for human-driven extinction of island fauna. Despite this general pattern, two recent documented extinctions were mainland species. The Carolina Parakeet *(Conuropsis carolinensis)* was once abundant and widely distributed across the eastern United States. The last reliable reports of Carolina Parakeets date from the 1920s, with several hundred museum skins all that remain of the only true United States parrot. The Paradise

Parrot *(Psephotus pulcherrimus)* inhabited the Darling Downs of southeast Queensland, becoming rare not long after pastoralists moved into its habitat. Credible sightings petered out by the 1930s, though claims of its existence continue to be made and debated, reflecting a reluctance to accept the demise of this beautiful parrot.

The most intriguing of the extinct parrots was the Broad-billed Parrot *(Lophopsittacus mauritianus)* of Mauritius. The species is known from bones plus the drawings and reports of early travelers. No museum skin exists. It was a large parrot with an unusual crest and massive bill. The plumage is usually described as grey, though it's possible one or both sexes were multicolored. Like the co-occurring Dodo *(Raphus cucullatus)*, the Broad-billed Parrot is often described as flightless. However, early writings did not mention this fact and illustrations show them perched at the tops of trees. They were last reported on Mauritius in 1673–1675, a time when the human population was relatively small. Their extinction was most likely the result of predation by rats and other introduced predators. The Mauritius Grey Parrot *(Lophopsittacus bensoni)* is another bird for which no museum skin exists. Similar in size to the extant Mauritius Parakeet *(Psittacula eques)*, it's thought to have survived until the 1760s. The Mascarene Islands were also home to several other extinct parrots. The Mascarene Parrot *(Mascarinus mascarinus)* was known from the island of Reunion. Museum skins and a

Table 10.2. Extinct parrots

Common name	Scientific name
Norfolk Island Kaka	*Nestor productus*
Rodrigues Parrot	*Necropsittacus rodericanus*
Raiatea Parakeet	*Cyanoramphus ulietanus*
Black-fronted Parakeet	*Cyanoramphus zealandicus*
Paradise Parrot	*Psephotus pulcherrimus*
Mascarene Parrot	*Mascarinus mascarinus*
Seychelles Parakeet	*Psittacula wardi*
Newton's Parakeet	*Psittacula exsul*
Jamaican Red Macaw	*Ara gossei*
Dominican Green-and-yellow Macaw	*Ara atwoodi*
Jamaican Green-and-yellow Macaw	*Ara erythrocephala*
Lesser Antillean Macaw	*Ara guadeloupensis*
Cuban Macaw	*Ara tricolor*
Guadeloupe Parakeet	*Aratinga labati*
Carolina Parakeet	*Conuropsis carolinensis*
Mauritius Grey Parrot	*Lophopsittacus bensoni*
Broad-billed Parrot	*Lophopsittacus mauritianus*
Guadeloupe Amazon	*Amazona violacea*
Martinique Amazon	*Amazona martinicana*

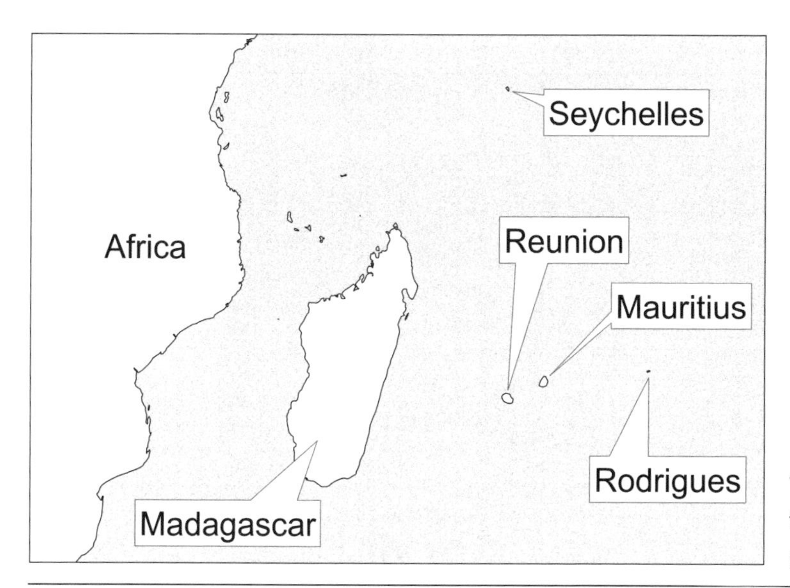

The Mascarene Islands were home to several now extinct species of parrot.

number of paintings reveal it to be a medium-sized brown parrot with a grey head, black face, and large red bill. Extinct in the wild by the end of the eighteenth century, captive birds survived in Europe until at least 1834. The island of Rodrigues supported two extinct parrots. The Rodrigues Parrot *(Necropsittacus rodericanus)* is known from early accounts and fossil bones. This large plain green parrot was last reported in 1761. Newton's Parakeet *(Psittacula exsul)* had the typical body shape of other *Psittacula*, but museum specimens have blue rather then green plumage. Based on contemporary descriptions, blue and green color morphs may have existed.

The Caribbean is another region where multiple parrot extinctions have been documented. The evidence for the existence of a number of these is less conclusive than for some of the Mascarene Islands parrots, with some authors suggesting they be described as hypothetical species. The Cuban Macaw *(Ara tricolor)* is an exception, with 19 specimens of this small red-blue-and-yellow macaw in existence. The last of these was collected in 1864, with hunting and the pet trade contributing to its decline. In *Extinct Birds* (2000), Errol Fuller considers the Jamaican Red Macaw *(A. gossei)*, described from a specimen procured in 1765 and no longer in existence, to be indistinguishable from the Cuban Macaw. The Lesser Antillean Macaw *(A. guadeloupensis)* was a large red-and-yellow macaw endemic to Guadeloupe and Martinique. While no specimens exist, a number of authors made detailed descriptions. The Jamaican Green-and-yellow Macaw *(A. erythrocephala)* and Dominican Green-and-yellow Macaw *(A. atwoodi)* have at times been considered a single species. Both are thought to have been extinct by the early nineteenth century. Given the current diversity of Amazon parrots in the Caribbean, it is not surprising that a number of species have been lost from the region. It's likely the islands of Guadeloupe

Table 10.3. Parrots listed by the IUCN as critically endangered (as of 13 March 2011)

Common name	Scientific name
Kakapo	*Strigops habroptila*
Yellow-crested Cockatoo	*Cacatua sulphurea*
Philippine Cockatoo	*Cacatua haematuropygia*
Blue-fronted Lorikeet	*Charmosyna toxopei*
New Caledonian Lorikeet	*Charmosyna diadema*
Red-throated Lorikeet	*Charmosyna amabilis*
Malherbe's Parakeet	*Cyanoramphus malherbi*
Orange-bellied Parrot	*Neophema chrysogaster*
Night Parrot	*Pezoporus occidentalis*
Glaucous Macaw	*Anodorhynchus glaucus*
Spix's Macaw	*Cyanopsitta spixii*
Blue-throated Macaw	*Ara glaucogularis*
Grey-breasted Parakeet	*Pyrrhura griseipectus*
Indigo-winged Parrot	*Hapalopsittaca fuertesi*
Puerto Rican Amazon	*Amazona vittata*

and Martinique each had its own species, *Amazona violacea* and *A. martinicana* respectively. The former species must have been a stunning bird, the feathers on its head, neck, and underparts being violet.

What is the world's rarest parrot?

Fifteen species of parrot are listed as critically endangered on the International Union for the Conservation of Nature (IUCN) "Red List." The recently described Western Ground Parrot *(Pezoporus flaviventris)* also falls into this category. Critically endangered species face "an extremely high risk of extinction in the wild." Many have small populations and thus qualify for consideration as the world's rarest parrot. However, determining which parrot deserves this dubious honor is difficult. Rarity and occupation of out-of-the-way habitats makes determining the size of populations problematic. Scientists often have to rely on unconfirmed reports of small numbers of birds at irregular intervals. A number of species listed as critically endangered have not been recorded for considerable periods of time and may be extinct in the wild.

The Spix's Macaw *(Cyanopsitta spixii)* is one of four large blue macaws native to South America. Its plight has received worldwide attention, and for a long time it was considered not only the world's rarest parrot but also the world's rarest bird. A single male was known to exist in the wild in 1990, keeping company with a female Blue-winged Macaw *(Primolius maracana)*. The fate of this individual was followed for a decade until its disappearance toward the end of 2000. Around 120 individuals are thought to exist in captivity, with 71 of these included in the official captive breeding program. There is a slim chance that a few birds may survive in the wild.

On the opposite side of the world, two island parrots survive as tiny populations. The New Caledonian Lorikeet *(Charmosyna diadema)* has been occasionally sighted over the last 100 years, but you have to go back to 1913 for the last confirmed record. The Red-throated Lorikeet *(C. amabilis)* occurs on a number of Fiji islands. It has been recorded more regularly than its New Caledonian relative, though sightings have declined in recent decades. Populations of both species are estimated to number fewer than 50 individuals, and their future appears bleak in the absence of concerted conservation action.

Will parrots be affected by global warming?

Many parrots will be negatively impacted by global warming. The impact of drought upon a population of Glossy Cockatoos *(Calyptorhynchus lathami)* in central New South Wales provides an insight into the likely impact of global warming on this and other parrot species. Glossy Cockatoos are dietary specialist and within Goonoo National Park feed on the cones of two shrubby sheoaks *(Allocasuarina)*. Birds prefer to feed on young cones, the availability of which depends on rainfall in the preceding year. A wet year produces a bumper seed crop the following year, which results in most Glossy Cockatoo pairs breeding. Conversely, drought years mean no young cones are available the following year, and few if any cockatoos attempt to breed. Between these two extremes, Glossy Cockatoo productivity in a given year reflects rainfall in the previous year. Historically, the number of Glossy Cockatoos would have fluctuated in response to climatic conditions. Any increase in aridity as a result of global warming will reduce seed production in sheoaks, causing Glossy Cockatoo populations to oscillate more strongly than they do at present. Populations will decline when the interval between periods of food shortage is insufficient to allow their complete recovery.

Higher temperatures and more frequent or intense droughts as a result of global warming will result in larger and more intense fires. These can eliminate Glossy Cockatoo food and nest resources over large areas. Following one of the driest years on record, a fire within Goonoo National Park destroyed approximately 50% of Glossy Cockatoo foraging habitat. This high-intensity fire destroyed all 24 known nest trees within the fire area, with these hollowed out forest giants burning fiercely once they had caught alight. Before-and-after surveys of large old trees revealed the majority had been destroyed in the fire, a catastrophic loss that will take hundreds of years to replace. Food resources would normally recover relatively quickly, sheoak seed being released from protective woody cones into the ash bed following passage of the fire. However, years of drought meant the quantity of seed stored in the canopy was relatively small and may be insufficient in some areas to ensure adequate regeneration. Regenerating sheoak stands can be eliminated if they are burnt before they have had time to mature and establish a canopy seed bank. Ongoing drought and high summer temperatures may result in the same country being burnt at relatively frequent intervals. These altered fire regimes have the potential to cause a shift in vegetation communities, with negative consequences for cockatoos dependent on them.

Why do people hunt parrots?

Birds were an important source of protein for traditional Maori. Kaka *(Nestor meridionalis)* were speared or snared, hunters working from a platform constructed in the canopy. Spears used for canopy hunting were six to eleven meters in length, with a barbed point made from bone, stone, or wood. Snares comprised a horizontal wooden perch over which was draped a single looped snare. Birds landing on the perch were caught by the feet when the hunter pulled a cord attached to the snare. A pet Kaka was sometimes used as a decoy. Trees on which birds were speared were referred to as *kaihua*, while snare trees were referred to as *tutu*. Trees regularly visited by parrots were a prized resource and some were individually named. Ceremonies were performed to increase the attractiveness to birds of particular trees, while the repetition of charms when setting snares helped ensure a good catch. Spiritual restrictions might be placed on the hunting of birds in parts of the forest or in particular seasons. If not eaten straight away, birds were preserved in their own fat. After being plucked and cleaned they were cooked on a spit, the dripping fat being collected in a wooden trough. The cooked birds were placed in containers and the reheated fat poured over them. Food was secured in storehouses, some of which were elaborately carved. These were usually elevated to prevent rats gaining entry.

The Maori incorporated parrot feathers into their garments. They wore cloaks woven from flax fiber, some of which had feathers scattered across the surface or woven into the borders. The production of cloaks completely covered in feathers did not commence until the second half of the nineteenth century, after the arrival of Europeans. These could take up to eight months to construct and were the preserve of high ranking individuals. Elsdon Best reported that cloaks covered with red Kaka feathers were especially prized. In the Papua New Guinea Highlands, bird plumes are also used decoratively. Christopher Healy reported on bird plumes contained

"Man carrying pineapples with a woven basket and two parrots on a stick, Rabaul, New Guinea, ca. 1929" by Sarah Johnston Chinnery (1887–1970), National Library of Australia.

within the collections of Maring men. The most widely represented species was the Sulphur-crested Cockatoo *(Cacatua galerita)*. Plumes from the Papuan Lorikeet *(Charmosyna papou)* were also considered to be of great decorative value, while those of several other lorikeets and Pesquet's Parrots *(Psittrichas fulgidus)* were of lesser importance. In total, plumes from 16 species of parrots were recorded from decorative collections. These were used as decorations in dances, cockatoo feathers being incorporated into headdresses and the tail plumes of Papuan Lorikeets being worn through the nose. Parrot skins provided additional embellishment.

What impact does the wildlife trade have on parrot populations?

Removal of nestlings for the pet trade significantly reduces nest success in many parrot populations. One study of Grey Parrots *(Psittacus erithacus)* found that 100% of observed nests failed due to poaching, while another reported that humans removed nestlings from 42% of nests. Angélica Rodríguez Castillo and Jessica Eberhard studied reproduction in the Yellow-crowned Amazon *(Amazona ochrocephala)* in western Panama. They monitored 63 nesting attempts over two seasons, finding less than 13% of nests successfully fledged young. A few nests were predated by Boas

(Boa constrictor), but the main cause of nest failure was poaching. At one location in the northeastern Peruvian Amazon, poachers harvested 26% of Orange-winged Amazon *(Amazona amazonica)* nestlings and 29% of Blue-and-yellow Macaw *(Ara ararauna)* nestlings. The ability of populations to tolerate varying levels of nest take depends on the species and environmental conditions. High levels of poaching will lead to the decline of most populations, while moderate levels will negatively impact on large parrots with limited reproductive potential. Low levels of nest take can be a problem for populations that are already experiencing poor nest success due to other factors. Information is lacking on what constitutes a sustainable harvest rate for most species, though levels exceeding 20% are cause for significant concern.

Poachers obtain nestlings by climbing to the nest hollow, cutting a hole in the trunk, or chopping down the tree. The latter methods are destructive and may have long-term consequences for populations due to a reduction in the number of nest sites. Most poachers target fully feathered nestlings, though eggs and small young are sometimes taken. Free-flying birds may be trapped using mist nets, snares, or plant-based glues, birds being vulnerable when they congregate at water points or food resources. Birds taken from the wild are sometimes marketed by poachers, but are usually bought by middlemen or commercial dealers. Poachers receive a small percentage of a bird's retail value, but the activity can still prove lucrative. One Neotropical study found poachers earned the equivalent of one to six months' wages each breeding season. Poaching can be an important contributor to the economy of local communities when other sources of income are lacking. José González reported on poaching activity centered on a protected area in northeastern Peru. Nestlings from several parrot species were harvested annually from *Mauritia* palm swamps by poachers operating from a nearby village. Poaching was an important source of income during the flood season, when money earned from agriculture and fishing declined. In some years, nearly half of all households in a village of around 1,000 people were engaged in poaching.

The best available data on the parrot trade are provided by signatories to the Convention on International Trade in Endangered Species of Wild Fauna and Flora (CITES). CITES is aimed at ensuring the international trade in wildlife does not threaten the survival of species. With the exception of a few commonly kept species such as the Budgerigar *(Melopsittacus undulatus)* and Cockatiel *(Nymphicus hollandicus)*, all parrots are listed on the appendices of the convention. Appendix I includes species threatened with extinction that should not be traded, while Appendix II includes species whose trade should be regulated to prevent unsustainable exploitation. Regulation entails the issuing of import and export licenses and the moni-

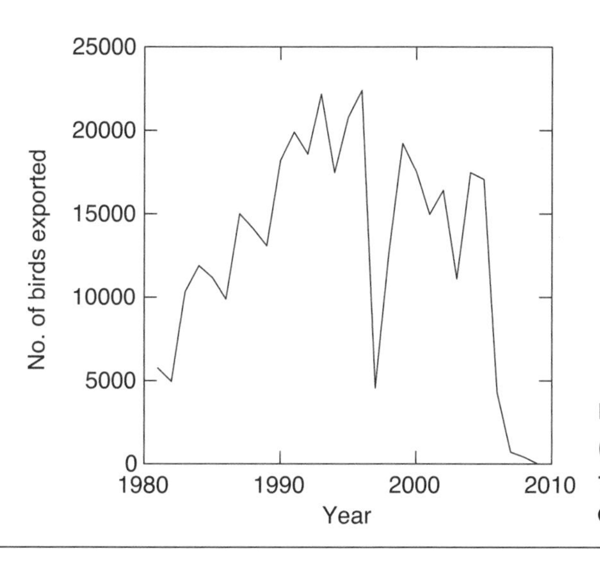

Reported legal export of Grey Parrots (*Psittacus erithacus*) from Cameroon 1981–2009. Data from UNEP-WCMC CITES Trade Database.

toring of wild populations. The numbers of parrots historically traded is staggeringly high, with reported figures greatly underestimating the number of birds removed from the wild, given mortality rates of 60% may be experienced prior to export. These figures also have no regard for birds that are illegally exported or traded within their country of origin. The numbers of traded parrots has declined sharply since the enactment of the U.S. Wild Bird Conservation Act (1992) effectively ended the shipment of wild-caught parrots into the United States. This legislation has proven effective in reducing demand and curtailing poaching activity. It is thought the 2007 decision of the European Union to end the importation of wild birds will have a similar effect. Many countries within the range of parrots have laws in place that prohibit or limit the taking of birds from the wild.

A recent study by Melvin Gastañaga and colleagues on the internal parrot trade in Peru highlighted the large number of birds involved. Over 12 months, they monitored wildlife markets in eight cities, recording the number of parrot species and individuals openly offered for sale. They found 34 species were traded, with 33 of these being native to Peru. The combined annual trade in the surveyed cities was estimated at 80,000 to 90,000 birds, with the vast majority thought to be sourced from the wild. With many major cities not included in the survey, the total number of parrots traded in Peru is likely to be much higher. While Peru allows for a small legal trade in several parrot species, the majority of the trade was illegal. A number of threatened parrots were present in the trade, raising concerns about its impact on wild populations. For example, 385 individuals of the endangered Grey-cheeked Parakeet *(Brotogeris pyrrhoptera)* were detected during the market surveys. A positive outcome of the investigation was that no parrots were offered for sale in Cuzco or Puerto Maldonado, the former an established tourism destination and the latter an emerging ecotourism center. It appears that tourism may serve to discourage the open trade in parrots.

Can captive breeding benefit wild populations of parrots?

Captive populations provide a safety net against the loss of species from the wild, but their ultimate purpose must be the provision of individuals for reintroduction programs. In this regard, it is important that founder birds capture the genetic diversity present within unique taxa (e.g., subspecies or population). Captive breeding programs supplement other conservation actions; they do not provide an excuse for failing to protect habitat or addressing other threats. Releases of captive-bred birds will only be successful when threats existing in the wild have been adequately mitigated. Releasing birds into protected areas offers the best chance of long-term success, though it is important that any potentially negative impacts on wild parrots and the environment are assessed. While there have been a relatively large number of parrot reintroductions, many of these have been small ad hoc releases or involved the translocation of wild birds from one site to another. Only a small number of threatened parrot recovery programs have adopted captive breeding and reintroduction as a recovery action.

The Orange-bellied Parrot *(Neophema chrysogaster)* is a critically endangered parrot that breeds in Tasmania and winters in southeastern Australia. As part of the recovery effort, an insurance captive population of around 170 birds has been established. Surplus birds from the captive population have been used to supplement the wild population. Releases have focused on reestablishing birds at a site formerly used for breeding, with 365 birds released between 1994 and 2009 at the selected site. Released birds have successfully bred, pairing with each other or wild birds. Captive-bred birds have successfully completed the annual migration undertaken by the species, including occasions where birds were not thought to be accompanied by wild parrots. For many years, the wild Orange-bellied Parrot population remained relatively stable at around 150 birds but in the last few years has plummeted to fewer than 50 individuals. This drop in numbers has been caused by a reduction in reproductive output, possibly due to a decline in habitat quality in the breeding or winter ranges. The captive population has now taken on added significance. The genetic base of the captive population, established from just six birds, will be broadened by the addition of nestlings or juveniles from the wild population.

The *Parrot Action Plan 2000–2004* highlights a number of limitations that have prevented the widespread adoption of captive breeding by recovery programs. Foremost among these are difficulties associated with releasing captive-bred birds into the wild. Parrots are highly social animals and rely on participation in flocks to find food and minimize predation risk. In addition, becoming proficient at processing food can take a number of years in some species. The young of many parrots spend an extended period of time learning the ropes, initially from their parents and later their flock mates. Providing social and learning opportunities for captive-bred birds can be problematic, highlighting the benefits of supplementing rather than reestablishing populations. The evolutionary pressures on captive birds are different from those existing in the wild. This raises the possibility of traits beneficial to wild birds being selected against by captive breeding, an issue that will become increasingly relevant as the proportion of wild-caught birds in captive populations decreases over time.

Another major issue facing captive breeding programs is the cost. The slow life histories of many parrots mean that producing large numbers of birds requires considerable investment in infrastructure and personnel. While avicultural techniques can be used to increase productivity (e.g., hand-raising of young), this can have undesirable consequences, including a loss of parenting skills in the captive population and the production of birds with a reduced probability of surviving in the wild. Captive parrot populations are susceptible to disease outbreaks. Disease poses a threat not only to birds held in breeding facilities, but also to wild populations into which captive birds are released. Disease risk can be minimized through hygiene and quarantine, but the construction of dedicated facilities within the natural range of the species is an added safeguard. The potential for birds held in captivity to transmit diseases to wild populations has led a number of authors to caution against the release of confiscated birds into the wild unless there is a clear conservation imperative and steps are taken to minimize the disease risk.

Chapter 11

Parrots in Stories and Literature

What roles do parrots play in religion and mythology?

Parrots are important birds to the Pueblo people of the American Southwest. Today, parrot feathers are incorporated into ritual objects used in traditional ceremonies. Macaw feathers are especially important, with the tail and wing feathers of Scarlet Macaws *(Ara macao)* highly prized. Archeological evidence suggests the use of parrots by Pueblo groups extends back at least 1,000 years. The ritualistic use of birds increased during the fourteenth century, which Suzanne Eckert associates with new ritual systems linked to the aggregation of Pueblo groups along major river systems. Archeological evidence for the importance of parrots comes from their depiction on pottery, portrayal on murals, and the recovery of physical remains of birds. Scarlet Macaws, Thick-billed Parrots *(Rhynchopsitta pachyrhyncha)*, and Amazon parrots have been described from pottery vessels. A kiva mural at the Pueblo village known as Pottery Mound shows a Thick-billed Parrot using its bill as an extra foot. As noted by Suzanne Eckert, this accurate depiction of parrot behavior indicates the artist was familiar with living birds. The most frequently encountered skeletal remains are of Scarlet Macaws, though Thick-billed Parrots and Military Macaws *(Ara militaris)* have also been uncovered. Parrot burials are usually associated with particular rooms thought to be associated with ceremonial activities. Macaws were typically 10–13 months old when they died, suggesting they were sacrificed in late winter or spring once their tail feathers were fully grown. Birds were plucked and may have had their left wing removed prior to formal burial.

The region encompassing the Pueblos was largely devoid of parrots, though the Thick-billed Parrot may have ranged into the southwestern United States. In particular, the favored Scarlet Macaw naturally occurred well to the south. This implies that macaws and their feathers were brought into the area. Based on pottery designs, juvenile macaws were transported in individual baskets carried on the backs of humans. The prehistoric settlement of Paquimé (Casas Grandes) in the northwest Mexican state of Chihuahua, itself outside the range of Scarlet Macaws, is thought to have been an important macaw breeding center or distribution point. Excavations at Paquimé have revealed the presence of adobe macaw pens and the remains of hundreds of parrots, the majority of them Scarlet Macaws. The macaw pens are characterized by an open top (thought to be covered in matting) and doors formed from a stone ring with pestle-like stone plug. Andrew Somerville and others have confirmed these pens were breeding rather than holding pens. Stable isotope analyses of carbon and oxygen in macaw bone carbonate showed birds were fed on plant foods unavailable to wild macaws and spent their entire lives at Paquimé. Today, Pueblo groups in the American Southwest obtain their feathers from captive birds. The Feather Distribution Project receives donated parrot feathers from zoos, breeders, and individuals across the United States and distributes these free of charge to Pueblo communities. Over the past 30 years, approximately three million parrot feathers have been provided for use in traditional cultural activities.

Parrots often feature in Australian Aboriginal stories. In a review of more than 400 documented stories about birds, Sonia Tidemann and Tim Whiteside were able to identify 116 species of mainland birds. One of the most common birds in these stories was the Sulphur-crested Cockatoo *(Cacatua galerita)*. Another 15 species of parrot featured, with the Red-tailed Cockatoo *(Calyptorhynchus banksii)*, Galah *(Eolophus roseicapillus)*, and Red-winged Parrot *(Aprosmictus erythropterus)* mentioned more than eight times. Aboriginal stories have a variety of purposes. Some explain the origins of landscape features, others serve as morality tales. Many are a testament to the knowledge held by Aboriginal people on the behavior and ecology of species. My favorite stories are those detailing the way birds acquired their current form, in particular the color of their plumage. The following story comes from Jawoyn people living on their traditional lands near Katherine in the Northern Territory.

> Jawayak-wayak (Black-faced Cuckoo-shrike) had a sore foot because it had a boil on it. Wakwak (Torresian Crow) went and burst the boil and, as he did so, pus flew up and into his eyes so that now Wakwak has white eyes. After that, Jawayak-wayak went and killed a kangaroo; but the other birds carried it because he still had a sore foot. They took it back home and set up

Adobe macaw pens excavated at the prehistoric settlement of Paquimé, Chihuahua, Mexico. Andrew Somerville

> a roast to cook it. Jawayak-wayak told Detdet (Rainbow Lorikeet) to take a slice of meat from the kangaroo and then fly away. He put a big slice of meat still hot and a bit raw on his back. Juices from the meat ran around onto his chest which became a reddish color and is still like that to this day. He also told Weley (Red-winged Parrot) to take a piece and put it on his wing, which became reddish and stayed red until even now. (Wynjorroc et al. 2001, in Tidemann and Whiteside 2010)

Did any early explorers or naturalists mention parrots in their writings?

The world map produced by Gerard Mercator and published in 1569 includes an imagined southern land mass. Encompassed within the border of this continent is a drawing of parrots and a legend that has been translated as follows: "Psitacorum regio [Region of Parrots]—so named by the Portuguese because of the unheard of size of the birds at that place. Driven there when sailing to Calicut by a libeccio wind, they followed the coast of this land for 2000 miles without finding an end to it." Donald and Molly Trounson view this reference as evidence of the Portuguese visiting the

Table 11.1. Parrots appearing in documented Australian Aboriginal stories, in order of decreasing prevalence

Common name	Scientific name
Sulphur-crested Cockatoo	*Cacatua galerita*
Red-tailed Cockatoo	*Calyptorhynchus banksii*
Red-winged Parrot	*Aprosmictus erythropterus*
Galah	*Eolophus roseicapillus*
Northern Rosella	*Platycercus venustus*
Budgerigar	*Melopsittacus undulatus*
Rainbow Lorikeet	*Trichoglossus haematodus*
Pink Cockatoo	*Lophochroa leadbeateri*
Carnaby's Cockatoo	*Calyptorhynchus latirostris*
Australian Ringneck	*Barnardius zonarius*
Cockatiel	*Nymphicus hollandicus*
Ground Parrot	*Pezoporus wallicus*
Little Corella	*Cacatua sanguinea*
Mulga Parrot	*Psephotus varius*
Pale-headed Rosella	*Platycercus adscitus*
Western Rosella	*Platycercus icterotis*

Source: Tidemann and Whiteside (2010).

southwest corner of Western Australia late in the fifteenth century. They consider that the large birds referred to in the text are white-tailed black-cockatoos, either Baudin's Cockatoo *(Calyptorhynchus baudinii)* or Carnaby's Cockatoo *(C. latirostris)*. Other European ships visited the Western Australian coastline in the following centuries, with expedition members noting the presence of parrots in their journals. In 1699, the English explorer William Damper reported seeing "a sort of white parrot, which flew a great many together." This sighting was made at Rosemary Island in the Dampier Archipelago and refers to the Little Corella *(Cacatua sanguinea)*. In 1801, the French expedition led by Nicolas Baudin called into Geographe Bay, named for one of the expedition's ships. Shore parties reported seeing "large black birds," subsequently named in honor of Baudin by Edward Lear in his 1832 publication *Illustrations of the Family of Psittacidae, or Parrots*.

While the west coast of Australia was still being explored, the east coast was being settled. Settlement of the coast was followed by exploration of the interior. Major Thomas Mitchell undertook a number of expeditions, documenting these in *Three Expeditions into the Interior of Eastern Australia*. He recorded Aboriginal people using nets to catch parrots: "The native knows well the alleys green through which at twilight the thirsty pigeons and parrots rush toward the water; and there, with a smaller net hung up,

Engraving of Nicolas Baudin by Conrad Westermayer (1765–1834), National Library of Australia.

he sits down and makes a fire ready to roast the birds which may fall into his snare." Mitchell was a keen observer of the local wildlife. Exploring the Darling River in 1835, his party came across a new species of parrot "having scarlet feathers on the breast, those on the head and wings being tinged a beautiful blue, and on the back etc. a dark brownish green." Specimens were collected and forwarded to John Gould, who described them at a meeting of the Zoological Society of London in 1837 as *Platycercus haematogaster*, a species we today know as the Bluebonnet *(Northiella haematogaster)*. Gould also received several additional new parrots from Mitchell's and other expeditions. These he subsequently named and included in his mammoth *The Birds of Australia* (1840–1848) and its supplement (1851–1869).

Alfred Russel Wallace was the foremost field naturalist of his day and his book *The Malay Archipelago*, published in 1869, contains many detailed references to the parrot fauna of the region. He observed the Salmon-crested Cockatoo *(Cacatua moluccensis)* in the Moluccas, noting the species was "commonly seen alive in Europe." The Chattering Lory *(Lorius garrulous)* was succinctly and accurately described as a "handsome red lory with green wings and a yellow spot in the back," while the Red-cheeked Parrot *(Geoffroyus geoffroyi)* was more poetically painted as "a green parrot with a red bill and head, which colour shaded on the crown into azure blue, and thence into verditer blue and the green of the back." Wallace was par-

"Cockatoo from the Interior of Australia" by Sir Thomas Mitchell (1792–1855), National Library of Australia. This watercolor of a Pink Cockatoo *(Lophochroa leadbeateri)* was reproduced in Vol. 2 of Mitchell's *Three Expeditions into the Interior of Eastern Australia.* The species is commonly known as "Major Mitchell's Cockatoo."

ticularly taken with the parrots he encountered in New Guinea: "Among its thirty species of parrots are the Great Black Cockatoo, and the little rigid-tailed Nasiterna [pygmy-parrot], the giant and the dwarf of the whole tribe. The bare-headed Dasyptilus [Pesquet's Parrot] is one of the most singular parrots known; while the beautiful little long-tailed Charmosyna, and the great variety of gorgeously-coloured lories, have no parallels elsewhere."

On the Aru Islands off the coast of New Guinea, Wallace obtained his first specimen of the Palm Cockatoo *(Probosciger aterrimus)*, which he described in great detail (an excellent block print accompanies his text). In what are some of the earliest published reports on tool use in birds, Wallace notes that a foraging Palm Cockatoo will "take hold of the nut with its foot, and biting off a piece of leaf retains it in the deep notch of the upper mandible, and again seizing the nut, which is prevented from slipping by the elastic tissue of the leaf, fixes the lower edge of the lower mandible in the notch, and by a powerful nip breaks off a piece of the shell." His skill as an ecologist is on display when he points out that "every detail of form and structure in the extraordinary bill of this bird seems to have its use, and we may easily conceive that the black cockatoos have maintained themselves in competition with their more active and more numerous white allies, by their power of existing on a kind of food which no other bird is able to extract from its stony shell."

What roles do parrots play in popular culture?

Parrots feature in a number of television comedies. Their role is to absorb the words and phrases of surrounding humans and repeat them at inopportune times, revealing secrets and causing embarrassment. In an episode from the first season of *Get Smart*, the Chief sends an Amazon parrot undercover to obtain information on a KAOS operation run from an animal spa. The Chief notes the parrot is "an expert at imitating voices" and "repeats everything he's heard." Unfortunately, the parrot reveals to KAOS that Max is a CONTROL agent, resulting in Max and 99 being captured. In an episode from the fourth season of *Frasier*, Niles moves into a new apartment and obtains a white-cockatoo as a replacement pet for his dog. Anxious to be accepted by his neighbors, he invites them to a dinner party. As the guests begin to arrive, the cockatoo gets a fright and cannot be dislodged from its perch atop Niles' head. Embarrassed to appear before his guests wearing a parrot, Niles spends much of the evening in the kitchen, where he and Frasier exchange uncomplimentary comments about the guests. When Niles eventually emerges with parrot in situ, his guests are very understanding and put him at ease. They are repaid for this generosity by the cockatoo repeating comments heard earlier in the kitchen, the bird mixing up some of the words to make things sound even worse. Affronted, the dinner guests leave.

A number of police procedural dramas have incorporated topical parrot issues into their storylines, including the illegal trade in parrots and parrots as invasive species. An episode of the tenth season of *Law & Order: Special Victims Unit* dealt with wildlife smuggling. Detective Olivia Benson recovers a dead parrot from a handbag found next to the body of a young woman. The parrot is identified as a Spix's Macaw *(Cyanopsitta spixii)* by Medical Examiner Melinda Warner, who shows a photo of a living bird and provides information on the species' conservation status. The murder is linked to an animal smuggling operation, which Detective Elliot Stabler attempts to infiltrate by posing as a corrupt customs officer. On one of Elliot's visits to the smugglers' warehouse, we are shown cages containing a Salmon-crested Cockatoo *(Cacatua moluccensis)* and a Scarlet Macaw *(Ara macao)*. An episode of the fourth season of *CSI: NY* has the team tracking down a serial killer, a cab driver who wraps his victims in tarpaulins prior to disposal. Avian fecal material is recovered from one of the tarpaulins, leading the team to a Monk Parakeet *(Myiopsitta monachus)* roost and the source of the tarpaulins. Dr. Sheldon Hawkes briefs Detective Stella Bonasera on the history of the birds, including their reputed escape from JFK in the 1960s. Hawkes notes the birds have "been multiplying in numbers and noise ever since."

In the 1990s there was a spate of family-friendly animal films, with a number starring parrots. These traded on the talking ability of parrots and their perceived wise-cracking habits. The hero of the movie *Paulie* is a Blue-crowned Parakeet *(Aratinga acuticaudata)* that travels across the United States searching for his best friend, a young girl. Along the way he meets up with a number of society's outsiders, who benefit from his help and teach him something in return. At one point, Paulie joins a taco-stand band comprising a Jandaya Parakeet *(A. jandaya)*, Nanday Parakeet *(Nandayus nenday)*, and Red-masked Parakeet *(Aratinga erythrogenys)*. Joseph Forshaw's and William Cooper's *Parrots of the World*, a very influential parrot book, is used to identify Paulie's species in one of the early scenes. The hero of the movie *The Real Macaw* is a talking Blue-and-yellow Macaw *(Ara ararauna)* named Mac. The film opens with pirates robbing an Amazon temple in 1850, capturing Mac in the process. The pirate ship sinks in the southwest Pacific, with the pirate captain surviving long enough to bury his treasure on a deserted island under Mac's watchful gaze. The movie then cuts to the present day, where Mac is keeping company with Sam and his laid-back, nature-loving Grandpa. With Grandpa short of cash and about to lose his house, Sam and Mac embark on an adventure to retrieve the pirate treasure.

What roles have parrots played in literature?

The book *The Thousand and One Nights* (also known as *The Arabian Nights*) features a story about a man, his wife, and a parrot. The woman's beauty was such that her husband felt obliged to stay at home lest she entertain a lover. Eventually obliged to make a journey, he purchased a parrot which he placed in his house as a spy. The parrot "was cunning and intelligent, and remembered whatever she heard." On his return, the man questioned the parrot about his wife's conduct and was informed she had been visited by her lover each night he was away. The wife learned of this conversation and the next night her husband was absent, she tricked the parrot into thinking there had been a storm. The parrot reported the storm to the husband, but he thought the bird was lying as the night had been calm. He killed the parrot in a rage, but later learned the truth and regretted his hasty actions. This ancient story has a modern-day parallel. In 2006, BBC NEWS reported that a parrot owner had been alerted to his girlfriend's infidelity when his Grey Parrot *(Psittacus erithacus)* began saying "I love you Gary" and making smooching sounds whenever the name Gary was said on TV. Turned out his girlfriend had been hooking up with "Gary" in the apartment she shared with her boyfriend. Unfortunately, the parrot owner had to give his pet away, unable to stand the bird repeating the name of his former girlfriend's lover.

The association between pirates and parrots owes much to Robert Louis Stevenson's *Treasure Island* (1882). In this classic pirate story, Long John Silver owns a parrot called Captain Flint. Silver suggests the bird may be two hundred years old and notes she has witnessed much wickedness in that time. The parrot is known for its swearing and repetition of the phrase "Pieces of eight!" The bird is incapable of moderating such behavior, bound to swear no matter the audience. Despite these annoying habits, Silver considers her "a handsome craft." The parrot is caged when first introduced, but is later observed on Silver's shoulder. This change in status appears to reflect the emergence of Silver's true character, his earlier obsequious behavior being dispensed with to reveal a ruthless buccaneer. Cap'n Flint is generally seen as a symbol of pirate greed and inhumanity, becoming the template for later literary sea-faring parrots. One such bird was Ginger, a parrot appearing in Lucy Montgomery's *Anne of Avonlea* (1909). Ginger belonged to Mr. Harrison, a new arrival to Avonlea and the first person in the area to keep a parrot. The parrot "swore terribly" and was considered an "unholy bird." It had formerly belonged to Mr. Harrison's brother, a sailor. Mr. Harrison confided in Anne Shirley that while there was nothing he hated more than profanity in a human being, "in a parrot, that's just repeating what it's heard with no more understanding of it than I'd have of Chinese, allowances might be made."

Mark Bittner's book *The Wild Parrots of Telegraph Hill* (2004) tells the story of the six years he spent with a flock of Red-masked Parakeets *(Aratinga erythrogenys)* in San Francisco. The book refers to the parakeets as Cherry-headed Conures, another of their common names. Bittner's observations on the behavior of individual birds are interwoven with the details of his personal life. When he first met the flock, Bittner was searching for a girlfriend, a career, and a deeper understanding of nature. Serendipitously, his love for the birds delivered him all these things. Documentary filmmaker Judy Irving filmed Bittner's last year with the flock, resulting in the award-winning *The Wild Parrots of Telegraph Hill.* The parakeets are beautifully shot on 16-mm film, often against a backdrop of San Francisco landmarks. While not a typical wildlife documentary, there is plenty of information for people interested in parrot biology and behavior. Parrots are shown feeding and breeding in the city's parks and gardens. The stories of individual flock members provide cinematic drama, with a Mitred Parakeet *(A. mitrata)* and Blue-crowned Parakeet *(A. acuticaudata)* starring alongside

Mark Bittner feeding the parrots.

Mark Bittner

Sam and Kristine. Mark Bittner

the Red-masked Parakeets. Bittner's voiceover takes us through his experiences with the flock and how these have led him to the conclusion that a parrot's concerns are not so different from those of humans and that "all life is one whole."

How long have humans kept and bred parrots?

The earliest record of humans keeping parrots is contained within the works of Ctesias, a Greek physician who served in the court of Artaxerxes II of Persia around 398 BC. On his return to Greece, he wrote a number of books on the Persian Empire and India. *Indika* reported what he had learned of India from travelers visiting the Persian court. In among descriptions of mythical peoples and fantastical beasts, it contains a relatively detailed account of a parrot. Andrew Nichols provides the following translation: "There is a bird called the *bittakos* which has a human voice, is capable of speech, and grows to the size of a falcon. It has a crimson face and a black beard and is dark blue as far as the neck . . . like cinnabar. It can converse like a human in Indian but if taught Greek, it can also speak Greek." Nichols notes that cinnabar is used to describe anything brilliant red. It is thought Ctesias was describing a Plum-headed Parakeet *(Psittacula cyanocephala).*

In *Elephant Slaves & Pampered Parrots* (2002), Louise Robbins documents the fascination eighteenth-century Parisians had with exotic animals, reflecting a growing interest in natural history. In response to this curiosity, large numbers of parrots began to arrive in France aboard ships returning from overseas ports. African parrots would often make the journey via the Caribbean, being embarked on ships transporting slaves to sugar plantations in the colonies. Much of the trade was private, sailors purchasing several birds and attempting to keep them alive for sale on their return to France. This was not necessarily an easy task, parrots featuring on the menu when a ship's company ran short of food. French ports were said to be alive with parrots, with birds sold to Parisian merchants for sale through shops specializing in birds and small animals. From these shops they made their way into households as family pets. The Grey Parrot *(Psittacus erithacus)*, known as the Jaco, was a popular species because of its talking ability. Robbins tracked the rising popularity and increased availability of parrots through the Paris newssheet, scouring the for-sale and lost-and-found columns for references to parrots and other pets. She found that beginning in the 1770s they were regular features, with one listing reading: "Young and beautiful amazon PARROT, very tame and beginning to speak well: 96 liv. Inquire of the porter of M. le chevalier *du Chaylar*, rue des petits Augustins."

Theodore Roosevelt, the twenty-sixth president of the United States, is

"Teddy, Jr. and 'Eli Yale,' 1902," by Frances Benjamin Johnston (1864–1952), Library of Congress.

known to have kept a Hyacinth Macaw *(Anodorhynchus hyacinthinus)* at the White House. In a letter thanking a young correspondent for her birthday wishes, Roosevelt included a drawing of the macaw and wrote, "the bird lives in the greenhouse, and is very friendly." The Library of Congress holds a photo of the bird, Eli Yale, perched on the hand of Roosevelt's son, Teddy Jr. Writing to author Joel Chandler Harris, Roosevelt commented that the bird had "a bill that I think could bite through boiler plate, who crawls all over Ted, and whom I view with dark suspicion." Roosevelt would later encounter Hyacinth Macaws in the wild during an expedition to Brazil in 1913–1914. These encounters were documented in *Through the Brazilian Wilderness* (1914): "On one ride we passed a clump of palms which were fairly ablaze with color. There were magnificent hyacinth macaws; green parrots with red splashes; toucans with varied plumage, black, white, red, yellow; green jacamars; flaming orioles and both blue and dark-red tanagers. It was an extraordinary collection. All were noisy. Perhaps there was a snake that had drawn them by its presence; but we could find no snake."

Chapter 12

"Parrotology"

Why study parrots?

We study parrots to obtain the information required to conserve them. Once, it was possible to create national parks large enough to ensure the long-term viability of natural communities without human input. The loss and fragmentation of habitat in recent decades means this is no longer the case. In addition, global warming makes establishing wilderness strongholds for species a redundant strategy. Accordingly, human intervention will often be required if biodiversity is to be maintained. A landscape approach to the conservation of species and their habitats is essential, but it needs to be informed by adequate information about the requirements of individual species. The collection of basic biological data is necessary if we are to ensure the ongoing availability of critical resources. The requirements of one species cannot necessarily be extrapolated from the findings of research undertaken on another species. If parrots are to be conserved, it is critical that research on their biology and behavior continues unabated in the coming decades. However, ecological research is not enough. Social and economic factors underpinning threats faced by parrots need to be understood and addressed if parrots are to have a long-term future.

Why bother to conserve parrots? At a purely practical level, many indigenous people include parrots in their diet. The extinction of local parrot populations from forests surrounding a settlement may result in seasonal food shortages. Parrots also have an important spiritual role in some cultures. A decline in parrot numbers can make obtaining the feathers or skins required for traditional ceremonies problematic, while the loss from the environment of a spiritually important species can create anxiety within

"Paradise Parrot *(Psephotus pulcherrimus)* photographed in the wild, Burnett River, Queensland, 1922" by Cyril Henry H. Jerrard (1889–1943), National Library of Australia. In this iconic shot, the male was photographed while pausing at the entrance to his nest.

communities. Humans are dependent on healthy ecosystems for a range of services, including the regulation of local and global climates. Our understanding of the roles parrots play within ecosystems is in its infancy. However, parrots are often a significant component of the vertebrate biomass (combined weight of all animals with backbones) and can be expected to exert some influence on the way communities function. Parrots provide pollination services to plants, limit the dominance of tree species by eating their seeds, and are important prey for predators. The critical role played by some parrots in the functioning of ecosystems may not become apparent until they disappear, providing an argument for a precautionary approach in relation to species extinction.

Many people consider that parrots, like all organisms, have an intrinsic worth separate to any benefit or service they provide to humans. Aldo Leopold captured this outlook in his classic *A Sand County Almanac* (1949), where he suggested that humans should "Quit thinking about decent land-use as solely an economic problem. Examine each question in terms of what is ethically and aesthetically right as well as what is economically expedient. A thing is right when it tends to preserve the integrity, beauty, and stability of the biotic community. It is wrong when it tends otherwise." Humans have a strong connection to nature regardless of whether they live in outback Australia or downtown New York. Research shows that people place a

“Man looking at Paradise Parrot *(Psephotus pulcherrimus)* nest in termite mound in the wild, Burnett River, Queensland, 1922,” by Cyril Henry H. Jerrard (1889–1943), National Library of Australia. This was the last confirmed nest of the species. Unfortunately, the eggs were infertile and the nest abandoned.

high value on the continued existence of wilderness. This value is independent of the likelihood of people ever visiting such areas, as evidenced by the popularity of documentaries featuring some out-of-the-way ecosystem or elusive species. Nevertheless, many people travel to see parrots, making a contribution to the economies of the countries and communities they visit.

Who studies parrots?

Biologists are hard at work around the planet improving our knowledge of the Psittaciformes. Basic information on the foraging ecology and breeding biology of species is being gathered, much of it collected by students undertaking postgraduate study. In the course of this work, information is obtained on threatening processes, and local communities are being engaged in conservation initiatives. Long-term studies are being run out of universities or supported by conservation organizations, with local and overseas volunteers often playing an important role. Collection of data over extended time frames is important due to the inherent variability of natural systems. In southeastern Australia, studies must run for several years if they are to encompass cyclical droughts. The data obtained from intensive study of wild parrots are being used to answer critical conservation questions, such as "How much land is required to conserve a parrot population?," "What types of habitat must be included in protected areas?," and "How should parrot habitat be managed?" Taxonomy plays an important role in parrot conservation by ensuring that resources are directed toward the highest priorities. The conservation of a small, isolated population of an abundant and widespread parrot may be a low priority, but if this population turns out to be a new species or unique taxon then its conservation becomes a high priority.

While parrots have long been recognized as social animals, we are only now beginning to appreciate the complex nature of parrot societies. There is a growing body of literature on the cognitive abilities of parrots and the importance of vocal communication in facilitating social networks. This research is inherently interesting and has made a significant contribution to our understanding of the social lives of animals and animal intelligence. Importantly, it can be used to improve the welfare of captive birds and inform conservation programs. The function of color in parrots is an expanding area of research. Early studies focused on how color influenced mate selection in captive birds, an avenue of investigation now being followed up in wild populations. Modern technologies have allowed researchers to see the world in the same way as parrots, revealing how "hidden colors" are used to signal condition and status without increasing predation risk. Studies of parrot color have the potential to assist with the noninvasive sexing of birds and the development of novel ways to monitor the health and status of populations. Research undertaken by zoos and aviculturists has provided information on how to successfully maintain captive populations capable of producing young birds fit for release into the wild.

Which species are best known?

The Australian cockatoos are one of the best-known groups of parrots. The impetus for early studies was the perception of cockatoos as significant agricultural pests. The Western Australian wheatbelt was the focus of this research, with scientists from the Commonwealth Scientific and Industrial Research Organization (CSIRO) undertaking landmark studies on Carnaby's Cockatoo *(Calyptorhynchus latirostris)*, Western Corellas *(Cacatua pastinator)*, Galahs *(Eolophus roseicapillus)*, and Pink Cockatoos *(Lophochroa lead-*

beateri). Later studies were undertaken on cockatoos considered pests in the eastern states, including the Long-billed Corella *(Cacatua tenuirostris)* and Sulphur-crested Cockatoo *(C. galerita).* In recent decades, cockatoo research in Australia has focused on threatened taxa, including studies of the Palm Cockatoo *(Probosciger aterrimus),* Baudin's Cockatoo *(Calyptorhynchus baudinii),* Carnaby's Cockatoo, Glossy Cockatoo *(C. lathami),* South-eastern Red-tailed Cockatoo *(C. banksii graptogyne),* Forest Red-tailed Cockatoo *(C. b. naso),* and Pink Cockatoo. However, gaps in our knowledge remain. We have little understanding of the foraging ecology or breeding biology of the declining Gang-gang Cockatoo *(Callocephalon fimbriatum),* while many common cockatoos remain unstudied throughout most of their range. Cockatoos occupying the islands to the north of Australia are poorly known. Surveys of a number of species have been completed, allowing conclusions to be drawn about numbers and habitat preferences. The Philippine Cockatoo *(Cacatua haematuropygia)* is the subject of a high-profile conservation program on the island of Palawan, while the breeding biology of the Palm Cockatoo has been studied in Papua New Guinea.

The Amazons are another group of relatively well-known parrots, due to the threatened status of many species. Around two-thirds of all Amazon parrots are threatened or near threatened, having restricted ranges and being negatively impacted by habitat loss and trapping for the bird trade. The best-known Amazons are those endemic to the Caribbean, all of which are threatened with extinction. A recovery program for the Puerto Rican Amazon *(Amazona vittata)* has been in place since the late 1960s, managed by the United States Fish and Wildlife Service and the Puerto Rican Department of Natural and Environmental Resources. This species is one of the few parrots to have its own scientific monograph, *The Parrots of Luquillo: Natural History and Conservation of the Puerto Rican Parrot* (1987). Other Amazons have benefited from the sustained attention of individual parrot biologists, an example being Katherine Renton's study of the Lilac-crowned Amazon *(A. finschi).* The large macaws are a group of Neotropical parrots for which there is reasonable knowledge. Many species are threatened and emblematic of the ecosystems in which they reside, generating interest among scientists and the wider community. Large parrots are often targeted by researchers because data collection is more straightforward than for smaller species. Parrots are notoriously difficult to study due to the difficulty of locating and following birds that are well camouflaged and capable of rapid flight. These problems are less apparent in large parrots. Also, local people are usually familiar with the habits of larger species and able to contribute to research programs.

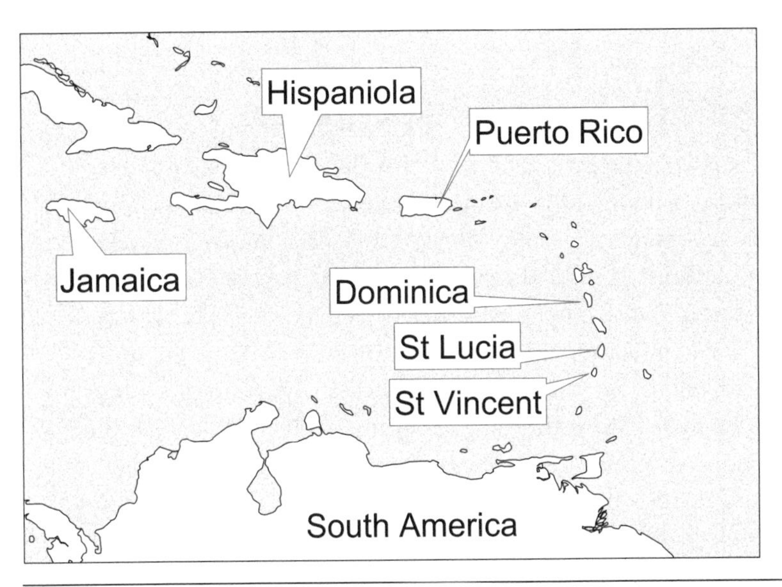

Many Caribbean islands host their own threatened Amazon parrot.

Which species are least known?

As a broad group, parrots inhabiting the Southeast Asian islands are the most poorly known of all the Psittaciformes. For many species and some genera, no scientific studies have been completed. The lack of information on lorikeets and racquet-tails is problematic given the high proportion of species that are threatened. There is little available published information on *Psittacula* parakeets, with a couple of notable exceptions. The endangered Mauritius Parakeet *(Psittacula eques)*, or Echo Parakeet, has been the subject of a long-running conservation program. The widely distributed Rose-ringed Parakeet *(P. krameri)* has attracted considerable attention because of its status as an invasive species, with the most detailed studies being undertaken outside the species' natural range. Many of the smaller Neotropical parrots are relatively unknown. The inconspicuous parrotlets are little studied, a concern given the limited distribution and threatened status of many species. The exception is the Green-rumped Parrotlet *(Forpus passerinus)*, one of the best-known Psittaciformes. Other Neotropical genera that would benefit from additional study include the *Pyrrhura* and *Hapalopsittaca*, both containing a high proportion of restricted range species. Overall, we have limited knowledge of most parrots. There is a need for a concerted research effort if we are to have sufficient information to develop effective conservation strategies for the many parrots that will require assistance in the coming decades.

How do scientists recognize individual parrots?

A few species of parrot have distinct markings that can be used to identify individuals within a population. Examples include the variable yellow blotches found on the heads of female Glossy Cockatoos *(Calyptorhynchus lathami)*, and the fine lines of feathers that decorate the faces of many *Ara* macaws. In most parrots, individuals can only be recognized if birds have been tagged. For birds generally, identification of individuals is achieved by placing bands on their legs. Steel bands stamped with a unique number allow birds to be identified if they are recaptured or found dead, while colored bands in unique combinations enable birds to be identified from a distance. The short tarsus of parrots makes it difficult to attach and subsequently read colored leg bands. To overcome this problem, scientists studying cockatoos in Western Australia in the 1970s used wing-tags. These were anodized aluminum discs stamped with a unique combination of letters. Wing-tags were found to result in increased mortality of Galahs *(Eolophus roseicapillus)* due to shooting by humans and increased predation of Carnaby's Cockatoos *(Calyptorhynchus latirostris)* by Wedge-tailed Eagles *(Aquila audax)*. A Galah that was caught and wing-tagged in June 1974 was brought into a veterinary clinic in September 2001, 27 years after it had been fitted with wing-tags. The tags were still legible and there was no evidence they had caused damage to the wings. The potential for increased mortality means wing-tags are no longer used to identify individual cockatoos. Recent attempts to monitor the movements of young Carnaby's Cockatoos by painting their tails have proved unsuccessful.

Radio tracking has been used extensively by parrot biologists to identify individual parrots and track their movements. It involves attaching a transmitter that emits a pulsed signal at a unique frequency. The researcher then uses a portable antenna and receiver to track the bird or triangulate its position. Transmitters can be attached in a variety of ways. Examining the impact of introduced mammal predators on Kaka *(Nestor meridionalis)*, Ron Moorhouse and colleagues mounted transmitters on the back of birds using harnesses. A weak link was incorporated into the harness, causing it to be shed after a set period of time. In all, transmitters were attached to 178 adult birds and 77 fledglings. The parrots quickly became used to the harness, with only a few removed by birds and no negative impacts detected. Studying habitat use in Mealy Amazons *(Amazona farinosa)*, Robin Bjork attached transmitters by means of a collar to 32 adult birds. The collars weighed 21 grams, which was 3% of body mass. The signal was initially located from vantage points within the study area (e.g., Mayan temple). Once located, the signal was tracked until the bird was visually located. An aircraft was used to find birds that could not be picked up by ground-based

Scarlet Macaw *(Ara macao)* fitted with satellite tracking collar, Tambopata Research Center, southeastern Peru. Matt Cameron

tracking. Technological advances mean scientists are now tracking parrots by satellite, with Donald Brightsmith employing satellite collars to monitor the movements of large macaws in southeastern Peru.

What is being done to save endangered parrot species?

The Mauritius Parakeet *(Psittacula eques)* was once considered the rarest parrot in the world, with the population in 1986 estimated to be 8–12 birds. The principal threat was loss of habitat, with just over 1% of the island's native vegetation remaining. Within the last vestiges of degraded habitat, parakeets struggled to find sufficient food and nest sites. They also had to cope with a number of predators and competitors, including Black Rats *(Rattus rattus)*, Crab-eating Macaques *(Macaca fascicularis)*, and Common Mynahs *(Acridotheres tristis)*. An introduced population of the closely related Rose-ringed Parakeet *(Psittacula krameri)* represented an added threat. Habitat restoration, predator control, and supplementary feeding have been important parts of the recovery effort, but there has been a heavy emphasis on the manipulation of breeding. Intensive monitoring of nests allowed researchers to remove poorly performing nestlings for hand-raising or fostering with other birds. Large broods of healthy chicks were downsized and small broods upsized to even out reproductive effort. These efforts resulted in the population increasing to the point that intensive management of breeding was reduced following the 2004–05 breeding season, with the species down-listed from Critically Endangered to Endangered in 2007. Supplementary feeding and the provision of nest boxes remain important parts of the recovery effort. The expression of Psittacine Beak and Feather Disease in the population since 2004 is cause for concern, with research into its prevalence, cause, and impacts currently underway. Following the fledging of 125 chicks in the 2010–11 breeding season, the population in April 2011 was estimated at 540 birds.

The Kangaroo Island Glossy Cockatoo *(Calyptorhynchus lathami halmaturinus)* numbered fewer than 150 birds in 1980, with surveys finding little evidence of successful breeding. In 1995, a recovery program commenced. Early research identified predation of eggs and nestlings by Common Brushtail Possums *(Trichosurus vulpecula)* as the principal cause of reproductive failure. Interference from other cockatoo species was also found, with populations of Galahs *(Eolophus roseicapillus)* and Little Corellas *(Cacatua sanguinea)* having increased due to altered environmental conditions. Honeybees occasionally usurped hollows previously used by Glossy Cockatoos. To protect nests from possums, trees were fitted with metal collars and canopy connections pruned. Targeted culling of nest competitors occurred, and bee swarms were removed from nesting areas. More than 80 artificial nest hollows were erected, with around half of these used by Glossy Cockatoos for nesting. The use of artificial hollows facilitates the protection and monitoring of nests as well as compensating for the loss of nesting habitat. These measures saw the population steadily increase from 195 birds in 1995, to a maximum of 360 birds in 2008. Accurate censuses have not been possible in recent years, though the population is thought to be stable. The carrying capacity of Kangaroo Island is thought to be around 650 birds. There is the potential for wildfire to eliminate large areas of habitat. To increase carrying capacity and reduce the potential impacts of wildfire, restoration of foraging and nesting habitat is being undertaken.

The Kakapo *(Strigops habroptila)* is perhaps the world's most familiar and popular parrot. The present-day recovery effort had its genesis in the discovery in 1977 of a remnant population of 100–200 birds on Stewart Island. A lack of mustelids on the island was thought to have contributed to the survival of these birds. In the 1980s, a decline in the Stewart Island population through cat predation led to all known birds being moved to islands free of mustelids and cats. Survival of adult birds on these offshore island sanctuaries was good, but limited reproduction occurred. Birds had trouble attaining breeding condition, while nesting attempts were disrupted by Pacific Rats *(Rattus exulans)*. The population reached a low of 51 birds in 1995, with only three young fledged between 1991 and 1995. The current focus of the recovery effort is Codfish Island, a 1,400-hectare nature reserve from which Pacific Rats were eliminated in the late 1990s.

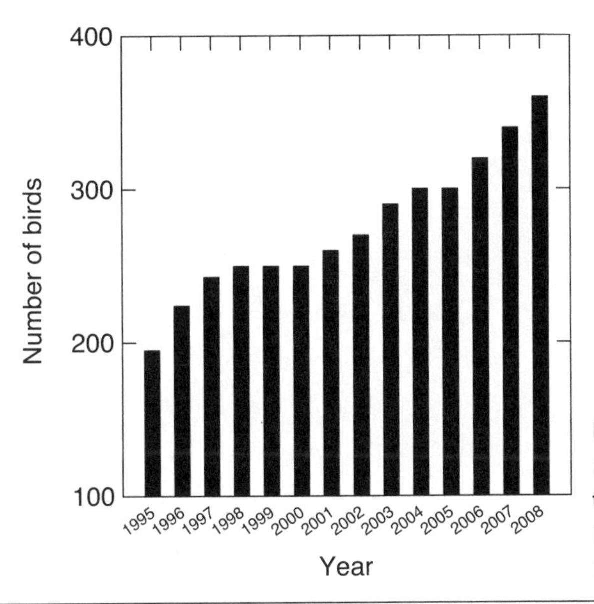

Kangaroo Island Glossy Cockatoo *(Calyptorhynchus lathami)* population estimate 1995–2008. Data from Department of Environment and Heritage, South Australia.

Birds roam free on the island, each carrying a radio transmitter that allows their activities to be monitored. All nests have their own human minder, who is responsible for keeping the chicks safe and warm in the absence of the female. Lack of food often leads to nest failure, so chicks that are not doing well are removed for hand-rearing. Considerable research has gone into developing a supplementary food that will enable breeding when natural foods are in short supply. This has been effective in improving egg production, though not all birds will utilize the supplements. Following a bumper breeding season in 2009 (33 chicks) and a good breeding season in 2011 (11 chicks), the population of Kakapo stood at 131 birds.

What can people do to help parrots?

Helping to ensure the persistence of parrot populations does not require a science degree. If you are fortunate enough to live in a region occupied by parrots, there is an opportunity to contribute your time to conservation efforts. A number of threatened species recovery programs in Australia rely on volunteer support. Much of our knowledge about the size and status of populations of Australian parrots has been derived from community-based surveys. The endangered Swift Parrot *(Lathamus discolor)* breeds in Tasmania, crossing Bass Strait to spend the winter on mainland south-eastern Australia. Volunteer survey weekends are held in May and August each year, providing valuable data on habitat use and movement patterns of birds on the mainland. These surveys have been running for 15 years, with around 300 volunteers completing more than 1,000 surveys annually. The South-eastern Red-tailed Cockatoo *(Calyptorhynchus banksii graptogyne)* is

Volunteers undertaking nest box maintenance prior to the Glossy Cockatoo *(Calyptorhynchus lathami)* breeding season on Kangaroo Island.
Michael Barth

restricted to southeast South Australia and southwest Victoria. Surveys of this endangered subspecies have been carried out each autumn for the last 15 years, with around 150 volunteers turning out to search the species' stringybark habitat. Community-based surveys are also used to monitor populations of other threatened Australian taxa, including the Orange-bellied Parrot *(Neophema chrysogaster)*, Carnaby's Cockatoo *(Calyptorhynchus latirostris)*, and several Glossy Cockatoo *(C. lathami)* populations.

A number of Australian parrot recovery projects provide opportunities for the local community to contribute directly to the conservation of threatened species. The Kangaroo Island Glossy Cockatoo *(C. l. halmaturinus)* recovery program has been running for 15 years. A feature of the recovery effort has been the level of community involvement. Up to 150 volunteers (or 3% of the island's population) turn out for the annual survey weekend. Volunteers also assist with the construction, erection, and maintenance of nest hollows, as well as locating and monitoring nests during the breeding season. The local State Emergency Service has trained nest-climbing teams in the use of single-rope techniques, while apiarists have worked to reduce the number of feral bee hives in cockatoo nest areas. The recovery program has been relatively well funded, allowing for the ongoing employment of a dedicated project officer. This has assisted in maintaining momentum within the community. The project officer produces the newsletter *Chewings*, named for the distinctive waste material produced by feeding Glossy Cockatoos, ensuring the island community is kept up to date with the recovery effort. The Kangaroo Island Glossy Cockatoo population has doubled over the life of the recovery program and is no longer considered critically endangered. Community support has played a large part in this conservation success story.

Don't despair if there is not a parrot conservation project underway in your backyard. A number of projects accept international volunteers. The Kakapo Recovery Program makes use of volunteers to provide supplementary food to birds over the summer months (i.e., carry heavy packs up steep hills), and to monitor nests in years when breeding occurs (i.e., stay up all night in atrocious weather). Seeing Kakapo is not guaranteed. The Tambopata Macaw Project, managed by Donald Brightsmith from the Schubot Center at Texas A&M University, undertakes research on macaws and other parrots inhabiting rainforest along the Tambopata River in the Peruvian Amazon. Much of the fieldwork is undertaken by teams of volunteers who have each made a six- to eight-week commitment to the project. As with the Kakapo program, the work is rewarding but physically demanding. The Tambopata Macaw Project also hosts Earthwatch Institute expeditions, providing less physically demanding opportunities for people to participate in the research project.

If fieldwork in remote areas is not your cup of tea, you can still help parrots by joining an organization dedicated to parrot conservation. BirdLife International is the body responsible for determining the conservation status of the world's birds for the International Union for the Conservation of Nature's (IUCN) Red List. It also plays an important role in identifying sites or issues that are a priority for conservation action. BirdLife International oversees the Important Bird Area Program. Important Bird Areas (IBAs) are sites supporting a large number of threatened or restricted-range species; they are priorities for conservation action. Responsibility for identification of IBAs usually falls to the BirdLife Partner in each country, typically a nongovernment organization committed to the conservation of birds. These partner organizations monitor the status of bird populations, lobby governments to protect birds and their habitats, and work with local communities to recover populations of threatened species. Membership in your local BirdLife Partner will help ensure that it remains a strong advocate for bird conservation and assist the implementation of BirdLife projects around the world.

The World Parrot Trust (WPT) aims to "work for the survival of all parrot species, and the welfare of every individual parrot." Established in 1989, the Trust has provided support for conservation and research programs benefiting around 50 species of parrot, including the Mauritius Parakeet *(Psittacula eques)*, Blue-throated Macaw *(Ara glaucogularis)*, and Thick-billed Parrot *(Rhynchopsitta pachyrhyncha)*. The WPT played an important role in the publication of the IUCN Parrot Action Plan, a document "providing researchers, managers, and local groups with practical recommendations for conducting conservation programs for threatened parrot species and populations endemic to their regions of the world." The WPT spearheaded a campaign to halt the importation of wild birds into the European Union. This objective was achieved with the adoption of Commission Regulation (EC) No 318/2007 of 23 March 2007 (applying from 1 July 2007), which found it appropriate to "limit imports of birds, other than poultry, only to birds bred in captivity." The impetus for this decision was the threat posed by avian influenza, but animal welfare issues also played a part. Membership in the WPT, or other organizations with a similar focus, provides an opportunity to learn about parrots and support people working to improve the lot of parrots in the wild and captivity.

Appendix

Common and Scientific Names of Living Parrots

Common name	Scientific name	Region
Kea	*Nestor notabilis*	Australasian
Kaka	*Nestor meridionalis*	Australasian
Kakapo	*Strigops habroptila*	Australasian
Pesquet's Parrot	*Psittrichas fulgidus*	Australasian
Vernal Hanging-parrot	*Loriculus vernalis*	Indomalayan
Sri Lanka Hanging-parrot	*Loriculus beryllinus*	Indomalayan
Philippine Hanging-parrot	*Loriculus philippensis*	Australasian
Blue-crowned Hanging-parrot	*Loriculus galgulus*	Indomalayan
Sulawesi Hanging-parrot	*Loriculus stigmatus*	Australasian
Moluccan Hanging-parrot	*Loriculus amabilis*	Australasian
Sula Hanging-parrot	*Loriculus sclateri*	Australasian
Sangihe Hanging-parrot	*Loriculus catamene*	Australasian
Orange-fronted Hanging-parrot	*Loriculus aurantiifrons*	Australasian
Green-fronted Hanging-parrot	*Loriculus tener*	Australasian
Red-billed Hanging-parrot	*Loriculus exilis*	Australasian
Yellow-throated Hanging-parrot	*Loriculus pusillus*	Indomalayan
Flores Hanging-parrot	*Loriculus flosculus*	Australasian
Yellow-capped Pygmy-parrot	*Micropsitta keiensis*	Australasian
Geelvink Pygmy-parrot	*Micropsitta geelvinkiana*	Australasian
Buff-faced Pygmy-parrot	*Micropsitta pusio*	Australasian
Meek's Pygmy-parrot	*Micropsitta meeki*	Australasian
Finsch's Pygmy-parrot	*Micropsitta finschii*	Australasian
Red-breasted Pygmy-parrot	*Micropsitta bruijnii*	Australasian
Palm Cockatoo	*Probosciger aterrimus*	Australasian
Baudin's Cockatoo	*Calyptorhynchus baudinii*	Australasian
Carnaby's Cockatoo	*Calyptorhynchus latirostris*	Australasian
Yellow-tailed Cockatoo	*Calyptorhynchus funereus*	Australasian
Red-tailed Cockatoo	*Calyptorhynchus banksii*	Australasian
Glossy Cockatoo	*Calyptorhynchus lathami*	Australasian
Gang-gang Cockatoo	*Callocephalon fimbriatum*	Australasian
Galah	*Eolophus roseicapillus*	Australasian
Pink Cockatoo	*Lophochroa leadbeateri*	Australasian
Yellow-crested Cockatoo	*Cacatua sulphurea*	Australasian

Sulphur-crested Cockatoo	*Cacatua galerita*	Australasian
Blue-eyed Cockatoo	*Cacatua ophthalmica*	Australasian
Salmon-crested Cockatoo	*Cacatua moluccensis*	Australasian
Umbrella Cockatoo	*Cacatua alba*	Australasian
Philippine Cockatoo	*Cacatua haematuropygia*	Australasian
Tanimbar Corella	*Cacatua goffiniana*	Australasian
Little Corella	*Cacatua sanguinea*	Australasian
Western Corella	*Cacatua pastinator*	Australasian
Long-billed Corella	*Cacatua tenuirostris*	Australasian
Solomons Corella	*Cacatua ducorpsii*	Australasian
Cockatiel	*Nymphicus hollandicus*	Australasian
Black Lory	*Chalcopsitta atra*	Australasian
Brown Lory	*Chalcopsitta duivenbodei*	Australasian
Yellow-streaked Lory	*Chalcopsitta sintillata*	Australasian
Cardinal Lory	*Chalcopsitta cardinalis*	Australasian
Red-and-blue Lory	*Eos histrio*	Australasian
Violet-necked Lory	*Eos squamata*	Australasian
Red Lory	*Eos bornea*	Australasian
Blue-streaked Lory	*Eos reticulata*	Australasian
Black-winged Lory	*Eos cyanogenia*	Australasian
Blue-eared Lory	*Eos semilarvata*	Australasian
Dusky Lory	*Pseudeos fuscata*	Australasian
Ornate Lorikeet	*Trichoglossus ornatus*	Australasian
Rainbow Lorikeet	*Trichoglossus haematodus*	Australasian
Olive-headed Lorikeet	*Trichoglossus euteles*	Australasian
Yellow-and-green Lorikeet	*Trichoglossus flavoviridis*	Australasian
Mindanao Lorikeet	*Trichoglossus johnstoniae*	Australasian
Pohnpei Lorikeet	*Trichoglossus rubiginosus*	Australasian
Scaly-breasted Lorikeet	*Trichoglossus chlorolepidotus*	Australasian
Varied Lorikeet	*Psitteuteles versicolor*	Australasian
Iris Lorikeet	*Psitteuteles iris*	Australasian
Goldie's Lorikeet	*Psitteuteles goldiei*	Australasian
Chattering Lory	*Lorius garrulus*	Australasian
Purple-naped Lory	*Lorius domicella*	Australasian
Black-capped Lory	*Lorius lory*	Australasian
Purple-bellied Lory	*Lorius hypoinochrous*	Australasian
White-naped Lory	*Lorius albidinucha*	Australasian
Yellow-bibbed Lory	*Lorius chlorocercus*	Australasian
Collared Lory	*Phigys solitarius*	Australasian
Blue-crowned Lorikeet	*Vini australis*	Australasian
Rimatara Lorikeet	*Vini kuhlii*	Australasian
Henderson Lorikeet	*Vini stepheni*	Australasian

Blue Lorikeet	*Vini peruviana*	Australasian
Ultramarine Lorikeet	*Vini ultramarina*	Australasian
Musk Lorikeet	*Glossopsitta concinna*	Australasian
Little Lorikeet	*Glossopsitta pusilla*	Australasian
Purple-crowned Lorikeet	*Glossopsitta porphyrocephala*	Australasian
Palm Lorikeet	*Charmosyna palmarum*	Australasian
Red-chinned Lorikeet	*Charmosyna rubrigularis*	Australasian
Meek's Lorikeet	*Charmosyna meeki*	Australasian
Blue-fronted Lorikeet	*Charmosyna toxopei*	Australasian
Striated Lorikeet	*Charmosyna multistriata*	Australasian
Pygmy Lorikeet	*Charmosyna wilhelminae*	Australasian
Red-fronted Lorikeet	*Charmosyna rubronotata*	Australasian
Red-flanked Lorikeet	*Charmosyna placentis*	Australasian
New Caledonian Lorikeet	*Charmosyna diadema*	Australasian
Red-throated Lorikeet	*Charmosyna amabilis*	Australasian
Duchess Lorikeet	*Charmosyna margarethae*	Australasian
Fairy Lorikeet	*Charmosyna pulchella*	Australasian
Josephine's Lorikeet	*Charmosyna josefinae*	Australasian
Papuan Lorikeet	*Charmosyna papou*	Australasian
Plum-faced Lorikeet	*Oreopsittacus arfaki*	Australasian
Yellow-billed Lorikeet	*Neopsittacus musschenbroekii*	Australasian
Orange-billed Lorikeet	*Neopsittacus pullicauda*	Australasian
Crimson Shining-parrot	*Prosopeia splendens*	Australasian
Masked Shining-parrot	*Prosopeia personata*	Australasian
Red Shining-parrot	*Prosopeia tabuensis*	Australasian
Horned Parakeet	*Eunymphicus cornutus*	Australasian
Uvea Parakeet	*Eunymphicus uvaeensis*	Australasian
Antipodes Parakeet	*Cyanoramphus unicolor*	Australasian
Norfolk Island Parakeet	*Cyanoramphus cookii*	Australasian
Red-crowned Parakeet	*Cyanoramphus novaezelandiae*	Australasian
New Caledonian Parakeet	*Cyanoramphus saisetti*	Australasian
Yellow-crowned Parakeet	*Cyanoramphus auriceps*	Australasian
Chatham Parakeet	*Cyanoramphus forbesi*	Australasian
Malherbe's Parakeet	*Cyanoramphus malherbi*	Australasian
Reischek's Parakeet*	*Cyanoramphus hochstetteri*	Australasian
Red-capped Parrot	*Purpureicephalus spurius*	Australasian
Australian Ringneck	*Barnardius zonarius*	Australasian
Green Rosella	*Platycercus caledonicus*	Australasian
Crimson Rosella	*Platycercus elegans*	Australasian
Northern Rosella	*Platycercus venustus*	Australasian
Pale-headed Rosella	*Platycercus adscitus*	Australasian
Eastern Rosella	*Platycercus eximius*	Australasian

Western Rosella	*Platycercus icterotis*	Australasian
Bluebonnet	*Northiella haematogaster*	Australasian
Red-rumped Parrot	*Psephotus haematonotus*	Australasian
Mulga Parrot	*Psephotus varius*	Australasian
Hooded Parrot	*Psephotus dissimilis*	Australasian
Golden-shouldered Parrot	*Psephotus chrysopterygius*	Australasian
Bourke's Parrot	*Neopsephotus bourkii*	Australasian
Blue-winged Parrot	*Neophema chrysostoma*	Australasian
Elegant Parrot	*Neophema elegans*	Australasian
Rock Parrot	*Neophema petrophila*	Australasian
Orange-bellied Parrot	*Neophema chrysogaster*	Australasian
Turquoise Parrot	*Neophema pulchella*	Australasian
Scarlet-chested Parrot	*Neophema splendida*	Australasian
Swift Parrot	*Lathamus discolor*	Australasian
Budgerigar	*Melopsittacus undulatus*	Australasian
Ground Parrot	*Pezoporus wallicus*	Australasian
Western Ground Parrot*	*Pezoporus flaviventris*	Australasian
Night Parrot	*Pezoporus occidentalis*	Australasian
Brehm's Tiger-parrot	*Psittacella brehmii*	Australasian
Painted Tiger-parrot	*Psittacella picta*	Australasian
Modest Tiger-parrot	*Psittacella modesta*	Australasian
Madarasz's Tiger-parrot	*Psittacella madaraszi*	Australasian
Blue-rumped Parrot	*Psittinus cyanurus*	Indomalayan
Red-cheeked Parrot	*Geoffroyus geoffroyi*	Australasian
Blue-collared Parrot	*Geoffroyus simplex*	Australasian
Singing Parrot	*Geoffroyus heteroclitus*	Australasian
Montane Racquet-tail	*Prioniturus montanus*	Australasian
Mindanao Racquet-tail	*Prioniturus waterstradti*	Australasian
Blue-headed Racquet-tail	*Prioniturus platenae*	Australasian
Green Racquet-tail	*Prioniturus luconensis*	Australasian
Blue-crowned Racquet-tail	*Prioniturus discurus*	Australasian
Blue-winged Racquet-tail	*Prioniturus verticalis*	Australasian
Yellow-breasted Racquet-tail	*Prioniturus flavicans*	Australasian
Golden-mantled Racquet-tail	*Prioniturus platurus*	Australasian
Buru Racquet-tail	*Prioniturus mada*	Australasian
Great-billed Parrot	*Tanygnathus megalorynchos*	Australasian
Blue-naped Parrot	*Tanygnathus lucionensis*	Australasian
Blue-backed Parrot	*Tanygnathus sumatranus*	Australasian
Black-lored Parrot	*Tanygnathus gramineus*	Australasian
Eclectus Parrot	*Eclectus roratus*	Australasian
Australian King-parrot	*Alisterus scapularis*	Australasian
Moluccan King-parrot	*Alisterus amboinensis*	Australasian

Papuan King-parrot	*Alisterus chloropterus*	Australasian
Olive-shouldered Parrot	*Aprosmictus jonquillaceus*	Australasian
Red-winged Parrot	*Aprosmictus erythropterus*	Australasian
Superb Parrot	*Polytelis swainsonii*	Australasian
Regent Parrot	*Polytelis anthopeplus*	Australasian
Princess Parrot	*Polytelis alexandrae*	Australasian
Alexandrine Parakeet	*Psittacula eupatria*	Indomalayan
Rose-ringed Parakeet	*Psittacula krameri*	Indomalayan, Afrotropical
Mauritius Parakeet	*Psittacula eques*	Afrotropical
Slaty-headed Parakeet	*Psittacula himalayana*	Indomalayan
Grey-headed Parakeet	*Psittacula finschii*	Indomalayan
Plum-headed Parakeet	*Psittacula cyanocephala*	Indomalayan
Blossom-headed Parakeet	*Psittacula roseata*	Indomalayan
Malabar Parakeet	*Psittacula columboides*	Indomalayan
Emerald-collared Parakeet	*Psittacula calthropae*	Indomalayan
Derbyan Parakeet	*Psittacula derbiana*	Indomalayan
Red-breasted Parakeet	*Psittacula alexandri*	Indomalayan
Nicobar Parakeet	*Psittacula caniceps*	Indomalayan
Long-tailed Parakeet	*Psittacula longicauda*	Indomalayan
Grey-headed Lovebird	*Agapornis canus*	Afrotropical
Red-headed Lovebird	*Agapornis pullarius*	Afrotropical
Black-winged Lovebird	*Agapornis taranta*	Afrotropical
Black-collared Lovebird	*Agapornis swindernianus*	Afrotropical
Rosy-faced Lovebird	*Agapornis roseicollis*	Afrotropical
Fischer's Lovebird	*Agapornis fischeri*	Afrotropical
Yellow-collared Lovebird	*Agapornis personatus*	Afrotropical
Lilian's Lovebird	*Agapornis lilianae*	Afrotropical
Black-cheeked Lovebird	*Agapornis nigrigenis*	Afrotropical
Vasa Parrot	*Coracopsis vasa*	Afrotropical
Black Parrot	*Coracopsis nigra*	Afrotropical
Grey Parrot†	*Psittacus erithacus*	Afrotropical
Cape Parrot*	*Poicephalus robustus*	Afrotropical
Brown-necked Parrot	*Poicephalus fuscicollis*	Afrotropical
Red-fronted Parrot	*Poicephalus gulielmi*	Afrotropical
Senegal Parrot	*Poicephalus senegalus*	Afrotropical
Niam-niam Parrot	*Poicephalus crassus*	Afrotropical
Meyer's Parrot	*Poicephalus meyeri*	Afrotropical
Yellow-fronted Parrot	*Poicephalus flavifrons*	Afrotropical
Red-bellied Parrot	*Poicephalus rufiventris*	Afrotropical
Brown-headed Parrot	*Poicephalus cryptoxanthus*	Afrotropical
Rueppell's Parrot	*Poicephalus rueppellii*	Afrotropical

Hyacinth Macaw	*Anodorhynchus hyacinthinus*	Neotropical
Lear's Macaw	*Anodorhynchus leari*	Neotropical
Glaucous Macaw	*Anodorhynchus glaucus*	Neotropical
Spix's Macaw	*Cyanopsitta spixii*	Neotropical
Blue-and-yellow Macaw	*Ara ararauna*	Neotropical
Blue-throated Macaw	*Ara glaucogularis*	Neotropical
Military Macaw	*Ara militaris*	Neotropical
Great Green Macaw	*Ara ambiguus*	Neotropical
Scarlet Macaw	*Ara macao*	Neotropical
Red-and-green Macaw	*Ara chloropterus*	Neotropical
Red-fronted Macaw	*Ara rubrogenys*	Neotropical
Chestnut-fronted Macaw	*Ara severus*	Neotropical
Red-bellied Macaw	*Orthopsittaca manilata*	Neotropical
Blue-headed Macaw	*Primolius couloni*	Neotropical
Blue-winged Macaw	*Primolius maracana*	Neotropical
Yellow-collared Macaw	*Primolius auricollis*	Neotropical
Red-shouldered Macaw	*Diopsittaca nobilis*	Neotropical
Thick-billed Parrot	*Rhynchopsitta pachyrhyncha*	Neotropical
Maroon-fronted Parrot	*Rhynchopsitta terrisi*	Neotropical
Yellow-eared Parrot	*Ognorhynchus icterotis*	Neotropical
Golden Parakeet	*Guaruba guarouba*	Neotropical
Blue-crowned Parakeet	*Aratinga acuticaudata*	Neotropical
Green Parakeet	*Aratinga holochlora*	Neotropical
Socorro Parakeet	*Aratinga brevipes*	Neotropical
Red-throated Parakeet	*Aratinga rubritorquis*	Neotropical
Scarlet-fronted Parakeet	*Aratinga wagleri*	Neotropical
Mitred Parakeet	*Aratinga mitrata*	Neotropical
Red-masked Parakeet	*Aratinga erythrogenys*	Neotropical
Crimson-fronted Parakeet	*Aratinga finschi*	Neotropical
White-eyed Parakeet	*Aratinga leucophthalma*	Neotropical
Cuban Parakeet	*Aratinga euops*	Neotropical
Hispaniolan Parakeet	*Aratinga chloroptera*	Neotropical
Sun Parakeet	*Aratinga solstitialis*	Neotropical
Sulphur-breasted Parakeet*	*Aratinga pintoi*	Neotropical
Jandaya Parakeet	*Aratinga jandaya*	Neotropical
Golden-capped Parakeet	*Aratinga auricapillus*	Neotropical
Dusky-headed Parakeet	*Aratinga weddellii*	Neotropical
Olive-throated Parakeet	*Aratinga nana*	Neotropical
Orange-fronted Parakeet	*Aratinga canicularis*	Neotropical
Peach-fronted Parakeet	*Aratinga aurea*	Neotropical
Brown-throated Parakeet	*Aratinga pertinax*	Neotropical
Cactus Parakeet	*Aratinga cactorum*	Neotropical

Nanday Parakeet	*Nandayus nenday*	Neotropical
Golden-plumed Parakeet	*Leptosittaca branickii*	Neotropical
Burrowing Parrot	*Cyanoliseus patagonus*	Neotropical
Blue-throated Parakeet	*Pyrrhura cruentata*	Neotropical
Blaze-winged Parakeet	*Pyrrhura devillei*	Neotropical
Maroon-bellied Parakeet	*Pyrrhura frontalis*	Neotropical
Pearly Parakeet	*Pyrrhura lepida*	Neotropical
Crimson-bellied Parakeet	*Pyrrhura perlata*	Neotropical
Green-cheeked Parakeet	*Pyrrhura molinae*	Neotropical
Painted Parakeet	*Pyrrhura picta*	Neotropical
Maroon-faced Parakeet	*Pyrrhura leucotis*	Neotropical
Grey-breasted Parakeet	*Pyrrhura griseipectus*	Neotropical
Pfrimer's Parakeet	*Pyrrhura pfrimeri*	Neotropical
Santa Marta Parakeet	*Pyrrhura viridicata*	Neotropical
Fiery-shouldered Parakeet	*Pyrrhura egregia*	Neotropical
Maroon-tailed Parakeet	*Pyrrhura melanura*	Neotropical
El Oro Parakeet	*Pyrrhura orcesi*	Neotropical
Black-capped Parakeet	*Pyrrhura rupicola*	Neotropical
White-necked Parakeet	*Pyrrhura albipectus*	Neotropical
Flame-winged Parakeet	*Pyrrhura calliptera*	Neotropical
Red-eared Parakeet	*Pyrrhura hoematotis*	Neotropical
Rose-headed Parakeet	*Pyrrhura rhodocephala*	Neotropical
Sulphur-winged Parakeet	*Pyrrhura hoffmanni*	Neotropical
Austral Parakeet	*Enicognathus ferrugineus*	Neotropical
Slender-billed Parakeet	*Enicognathus leptorhynchus*	Neotropical
Monk Parakeet	*Myiopsitta monachus*	Neotropical
Grey-hooded Parakeet	*Psilopsiagon aymara*	Neotropical
Mountain Parakeet	*Psilopsiagon aurifrons*	Neotropical
Barred Parakeet	*Bolborhynchus lineola*	Neotropical
Andean Parakeet	*Bolborhynchus orbygnesius*	Neotropical
Rufous-fronted Parakeet	*Bolborhynchus ferrugineifrons*	Neotropical
Mexican Parrotlet	*Forpus cyanopygius*	Neotropical
Green-rumped Parrotlet	*Forpus passerinus*	Neotropical
Blue-winged Parrotlet	*Forpus xanthopterygius*	Neotropical
Spectacled Parrotlet	*Forpus conspicillatus*	Neotropical
Dusky-billed Parrotlet	*Forpus modestus*	Neotropical
Pacific Parrotlet	*Forpus coelestis*	Neotropical
Yellow-faced Parrotlet	*Forpus xanthops*	Neotropical
Plain Parakeet	*Brotogeris tirica*	Neotropical
White-winged Parakeet	*Brotogeris versicolurus*	Neotropical
Yellow-chevroned Parakeet	*Brotogeris chiriri*	Neotropical
Grey-cheeked Parakeet	*Brotogeris pyrrhoptera*	Neotropical

Orange-chinned Parakeet	*Brotogeris jugularis*	Neotropical
Cobalt-winged Parakeet	*Brotogeris cyanoptera*	Neotropical
Golden-winged Parakeet	*Brotogeris chrysoptera*	Neotropical
Tui Parakeet	*Brotogeris sanctithomae*	Neotropical
Tepui Parrotlet	*Nannopsittaca panychlora*	Neotropical
Amazonian Parrotlet	*Nannopsittaca dachilleae*	Neotropical
Lilac-tailed Parrotlet	*Touit batavicus*	Neotropical
Scarlet-shouldered Parrotlet	*Touit huetii*	Neotropical
Red-fronted Parrotlet	*Touit costaricensis*	Neotropical
Blue-fronted Parrotlet	*Touit dilectissimus*	Neotropical
Sapphire-rumped Parrotlet	*Touit purpuratus*	Neotropical
Brown-backed Parrotlet	*Touit melanonotus*	Neotropical
Golden-tailed Parrotlet	*Touit surdus*	Neotropical
Spot-winged Parrotlet	*Touit stictopterus*	Neotropical
Black-headed Parrot	*Pionites melanocephalus*	Neotropical
White-bellied Parrot	*Pionites leucogaster*	Neotropical
Brown-hooded Parrot	*Pyrilia haematotis*	Neotropical
Rose-faced Parrot	*Pyrilia pulchra*	Neotropical
Saffron-headed Parrot	*Pyrilia pyrilia*	Neotropical
Orange-cheeked Parrot	*Pyrilia barrabandi*	Neotropical
Caica Parrot	*Pyrilia caica*	Neotropical
Bald Parrot	*Pyrilia aurantiocephala*	Neotropical
Vulturine Parrot	*Pyrilia vulturina*	Neotropical
Black-winged Parrot	*Hapalopsittaca melanotis*	Neotropical
Rusty-faced Parrot	*Hapalopsittaca amazonina*	Neotropical
Indigo-winged Parrot	*Hapalopsittaca fuertesi*	Neotropical
Red-faced Parrot	*Hapalopsittaca pyrrhops*	Neotropical
Pileated Parrot	*Pionopsitta pileata*	Neotropical
Short-tailed Parrot	*Graydidascalus brachyurus*	Neotropical
Yellow-faced Amazon	*Alipiopsitta xanthops*	Neotropical
Blue-headed Parrot	*Pionus menstruus*	Neotropical
Red-billed Parrot	*Pionus sordidus*	Neotropical
Scaly-headed Parrot	*Pionus maximiliani*	Neotropical
Speckle-faced Parrot	*Pionus tumultuosus*	Neotropical
White-crowned Parrot	*Pionus senilis*	Neotropical
Bronze-winged Parrot	*Pionus chalcopterus*	Neotropical
Dusky Parrot	*Pionus fuscus*	Neotropical
Cuban Amazon	*Amazona leucocephala*	Neotropical
Yellow-billed Amazon	*Amazona collaria*	Neotropical
Hispaniolan Amazon	*Amazona ventralis*	Neotropical
White-fronted Amazon	*Amazona albifrons*	Neotropical
Yellow-lored Amazon	*Amazona xantholora*	Neotropical

Black-billed Amazon	*Amazona agilis*	Neotropical
Puerto Rican Amazon	*Amazona vittata*	Neotropical
Tucuman Amazon	*Amazona tucumana*	Neotropical
Red-spectacled Amazon	*Amazona pretrei*	Neotropical
Red-crowned Amazon	*Amazona viridigenalis*	Neotropical
Lilac-crowned Amazon	*Amazona finschi*	Neotropical
Red-lored Amazon	*Amazona autumnalis*	Neotropical
Blue-cheeked Amazon	*Amazona dufresniana*	Neotropical
Red-browed Amazon	*Amazona rhodocorytha*	Neotropical
Red-tailed Amazon	*Amazona brasiliensis*	Neotropical
Festive Amazon	*Amazona festiva*	Neotropical
Yellow-shouldered Amazon	*Amazona barbadensis*	Neotropical
Blue-fronted Amazon	*Amazona aestiva*	Neotropical
Yellow-headed Amazon	*Amazona oratrix*	Neotropical
Yellow-naped Amazon	*Amazona auropalliata*	Neotropical
Yellow-crowned Amazon	*Amazona ochrocephala*	Neotropical
Orange-winged Amazon	*Amazona amazonica*	Neotropical
Scaly-naped Amazon	*Amazona mercenaria*	Neotropical
Kawall's Amazon	*Amazona kawalli*	Neotropical
Mealy Amazon	*Amazona farinosa*	Neotropical
Vinaceous Amazon	*Amazona vinacea*	Neotropical
St Lucia Amazon	*Amazona versicolor*	Neotropical
Red-necked Amazon	*Amazona arausiaca*	Neotropical
St Vincent Amazon	*Amazona guildingii*	Neotropical
Imperial Amazon	*Amazona imperialis*	Neotropical
Red-fan Parrot	*Deroptyus accipitrinus*	Neotropical
Blue-bellied Parrot	*Triclaria malachitacea*	Neotropical
Orange-breasted Fig-parrot	*Cyclopsitta gulielmitertii*	Australasian
Double-eyed Fig-parrot	*Cyclopsitta diophthalma*	Australasian
Large Fig-parrot	*Psittaculirostris desmarestii*	Australasian
Edwards's Fig-parrot	*Psittaculirostris edwardsii*	Australasian
Salvadori's Fig-parrot	*Psittaculirostris salvadorii*	Australasian
Guaiabero	*Bolbopsittacus lunulatus*	Australasian

* Not recognized as a species in the BirdLife International "Checklist of the Birds of the World, version 4, 2011."
† The West African form of the Grey Parrot has recently been recognized as a separate species, the Timneh Grey Parrot *(Psittacus timneh)*.

Selected Bibliography

A complete bibliography may be found at the web page for this title at www.jhu.press.edu.

Aïn, S. A., N. Giret, M. Grand, M. Kreutzer, and D. Bovet. 2009. The discrimination of discrete and continuous amounts in African grey parrots *(Psittacus erithacus). Animal Cognition* 12:145–54.

Amuno, J. B., R. Massa, and C. Dranzoa. 2007. Abundance, movements and habitat use by African Grey Parrots *(Psittacus erithacus)* in Budongo and Mabira forest reserves, Uganda. *Ostrich* 78:225–31.

Arnold, K. E., I. P. F. Owens, and N. J. Marshall. 2002. Fluorescent signaling in parrots. *Science* 295:92.

Aumann, T. 2001. An intraspecific and interspecific comparison of raptor diets in the south-west of the Northern Territory, Australia. *Wildlife Research* 28:379–93.

Avery, M. L., C. A. Yoder, E. A. Tillman. 2008. Diazacon inhibits reproduction in invasive monk parakeet populations. *Journal of Wildlife Management* 72:1449–52.

Barrett, C. 1949. *Parrots of Australasia.* N. H. Seward Pty. Ltd., Melbourne.

Beckers, G. J. L., B. S. Nelson, and R. A. Suthers. 2004. Vocal-tract filtering by lingual articulation in a parrot. *Current Biology* 14:1592–97.

Beissinger, S. R., and J. R. Waltman. 1991. Extraordinary clutch size and hatching asynchrony of a Neotropical parrot. *Auk* 108:863–71.

Bell, H. L. 1982. A bird community of lowland rainforest in New Guinea. I. Composition and density of the avifauna. *Emu* 82:24–41.

Berg, M. L., and A. T. D. Bennett. 2010. The evolution of plumage colouration in parrots: A review. *Emu* 110:10–20.

Berkunsky, I., and J. C. Reboreda. 2009. Nest-site fidelity and cavity reoccupation by Blue-fronted Parrots *Amazona aestiva* in the dry Chaco of Argentina. *Ibis* 151:145–50.

Best, E. W. 1934. *The Maori as He Was: A Brief Account of Life as It Was in Pre-European Days.* Dominion Museum, Wellington.

Bittner, M. 2004. *The Wild Parrots of Telegraph Hill.* Three Rivers Press, New York.

Bjork, R. D. 2004. Delineating pattern and process in tropical lowlands: Mealy Parrot migration dynamics as a guide for regional conservation planning. PhD thesis, Oregon State University.

Boles, W. E. 1991. Glowing parrots—need for a study of hidden colors. *Birds International* 3:76–79.

———. 1993. A new cockatoo (Psittaciformes: Cacatuidae) from the Tertiary of Riversleigh, northwestern Queensland, and an evaluation of rostral characters in the systematics of parrots. *Ibis* 135:8–18.

Boles, W. E., N. W. Longmore, and M. C. Thompson. 1994. A recent specimen of the Night Parrot *Geopsittacus occidentalis. Emu* 94:37–40.

Bond, A. B., and J. Diamond. 2005. Geographic and ontogenetic variation in the contact calls of the kea *(Nestor notabilis). Behaviour* 142:1–20.

Bonebrake, T. C., and S. R. Beissinger. 2010. Predation and infanticide influence ideal free choice by a parrot occupying heterogeneous tropical habitats. *Oecologia* 163:385–93.

Boyer, J. S., L. L. Hass, M. H. Lurie, and D. T. Blumstein. 2006. Effect of visibility on time allocation and escape decisions in crimson rosellas. *Australian Journal of Zoology* 54:363–67.

Boyes, R. S., and M. R. Perrin. 2010. Aerial surveillance by a generalist seed predator: Food resource tracking by Meyer's parrot *Poicephalus meyeri* in the Okavango Delta, Botswana. *Journal of Tropical Ecology* 26:381–92.

Bradbury, J. W. 2003. Vocal communication in wild parrots. In *Animal Society Complexity: Intelligence, Culture, and Individualized Societies*, eds. F. B. M. de Waal and P. L. Tyack, 293–316, Harvard University Press, Cambridge.

Bradbury, J. W., K. A. Cortopassi, and J. R. Clemmons. 2001. Geographical variation in the contact calls of Orange-fronted Parakeets. *Auk* 118:958–72.

Brightsmith, D. J. 2005. Competition, predation and nest niche shifts among tropical cavity nesters: Ecological evidence. *Journal of Avian Biology* 36:74–83.

———. 2005. Parrot nesting in southeastern Peru: Seasonal patterns and keystone trees. *Wilson Bulletin* 117:296–305.

Brightsmith, D., and A. Bravo. 2006. Ecology and management of nesting blue-and-yellow macaws (*Ara ararauna*) in *Mauritia* palm swamps. *Biodiversity and Conservation* 15:4271–87.

Brown, E. D., and M. J. G. Hopkins. 1996. How New Guinea rainforest flower resources vary in time and space: Implications for nectarivorous birds. *Australian Journal of Ecology* 21:363–78.

Bucher, E. H. 1992. Neotropical parrots as agricultural pests. In *New World Parrots in Crisis: Solutions from Conservation Biology*, eds. S. R. Beissinger and N. F. R. Snyder, 201–19, Smithsonian Institution Press, Washington, DC.

Buckley, F. G. 1968. Behaviour of the Blue-crowned Hanging Parrot *Loriculus galgulus* with comparative notes on the Vernal Hanging Parrot *L. vernalis*. *Ibis* 110:145–64.

Budden, A. E., and S. R. Beissinger. 2009. Resource allocation varies with parental sex and brood size in the asynchronously hatching green-rumped parrotlet (*Forpus passerinus*). *Behavioural Ecology and Sociobiology* 63:637–47.

Buhrman-Deever, S. C., E. A. Hobson, and A. D. Hobson. 2008. Individual recognition and selective response to contact calls in foraging brown-throated conures, *Aratinga pertinax*. *Animal Behaviour* 76:1715–25.

Burtt, E. H., M. R. Schroeder, L. A. Smith, J. E. Sroka, and K. J. McGraw. 2011. Colourful parrot feathers resist bacterial degradation. *Biology Letters* 7:214–16.

Cameron, M. 2007. *Cockatoos*. CSIRO Publishing, Melbourne.

Cannon, C. E. 1984. Flock size of feeding eastern and pale-headed rosellas (Aves: Psittaciformes). *Australian Wildlife Research* 11:349–55.

Carwardine, M. 2009. *Last Chance to See: In the Footsteps of Douglas Adams*. Harper-Collins, London.

Cayley, N. W. 1935. *Budgerigars in Bush and Aviary*. 2nd. ed. Angus and Robertson, Sydney.

Chapman, T. F., and D. C. Paton. 2005. The glossy black-cockatoo (*Calyptorhynchus lathami halmaturinus*) spends little time and energy foraging on Kangaroo Island, South Australia. *Australian Journal of Zoology* 53:177–83.

Cheke, A. S. 1987. An ecological history of the Mascarene Islands, with particular reference to extinctions and introductions of land vertebrates. In *Studies of*

Mascarene Island Birds, ed. A. W. Diamond, 5–89, Cambridge University Press, Cambridge.

Collar, N. J. 1997. Family Psittacidae (Parrots). In *Handbook of the Birds of the World*, eds. J. del Hoyo, A. Elliot, and J. Sargatal, vol. 4, Sandgrouse to Cuckoos, 280–477, Lynx Edicions, Barcelona.

Cooney, S. J. N. 2009. Ecological associations of the Hooded Parrot *(Psephotus dissimilis)*. PhD thesis, Australian National University.

Cooper, C. E., P. C. Withers, P. R. Mawson, S. D. Bradshaw, J. Prince, and H. Robertson. 2002. Metabolic ecology of cockatoos in the south-west of Western Australia. *Australian Journal of Zoology* 50:67–76.

Corlett, R. T., and R. B. Primack. 2011. *Tropical Rain Forests: An Ecological and Biogeographical Comparison*. 2nd ed. Wiley-Blackwell, Chichester.

Courtney, J., and S. J. S. Debus. 2006. Breeding habits and conservation status of the Musk Lorikeet *Glossopsitta concinna* and Little Lorikeet *G. pusilla* in northern New South Wales. *Australian Field Ornithology* 23:109–24.

Diamond, J., and A. B. Bond. 1999. *Kea, Bird of Paradox: The Evolution and Behavior of a New Zealand Parrot*. University of California Press, Berkeley.

Diamond, J., D. Eason, C. Reid, and A. B. Bond. 2006. Social play in kakapo *(Strigops habroptilus)* with comparisons to kea *(Nestor notabilis)* and kaka *(Nestor meridionalis)*. *Behaviour* 143:1397–1423.

Díaz, S., and T. Kitzberger. 2006. High *Nothofagus* flower consumption and pollen emptying in the southern South American austral parakeet *(Enicognathus ferrugineus)*. *Austral Ecology* 31:759–66.

Eberhard, J. R. 1998. Breeding biology of the Monk Parakeet. *Wilson Bulletin* 110: 463–73.

———. 2002. Cavity adoption and the evolution of coloniality in cavity-nesting birds. *Condor* 104:240–47.

Eberhard, J. R., and E. Bermingham. 2005. Phylogeny and comparative biogeography of *Pionopsitta* parrots and *Pteroglossus* toucans. *Molecular Phylogenetics and Evolution* 36:288–304.

Eckert, S. L., and T. Clark. 2009. The ritual importance of birds in 14th-century central New Mexico. *Journal of Ethnobiology* 29:8–27.

Edwards, E. D., S. J. N. Cooney, P. D. Olsen, and S. T. Garnett. 2007. A new species of *Trisyntopa* Lower (Lepidoptera: Oecophoridae) associated with the nests of the hooded parrot *(Psephotus dissimilis*, Psittacidae) in the Northern Territory. *Australian Journal of Entomology* 46:276–80.

Ekstrom, J. M. M., T. Burke, L. Randrianaina, and T. R. Birkhead. 2007. Unusual sex roles in a highly promiscuous parrot: The Greater Vasa Parrot *Caracopsis vasa*. *Ibis* 149:313–20.

Fleming, T. H., C. F. Williams, F. J. Bonaccorso, and L. H. Herbst. 1985. Phenology and seed dispersal, and colonization in *Muntingia calabura*, a neotropical pioneer tree. *American Journal of Botany* 72:383–91.

Forshaw, J. M., and W. T. Cooper. 1989. *Parrots of the World*. 3rd ed. Lansdowne Editions, Sydney.

Forshaw, J. M., and F. Knight. 2006. *Parrots of the World: An Identification Guide*. Princeton University Press, Princeton.

Francisco, M. R., V. de O. Lunardi, and M. Galetti. 2002. Massive seed predation of *Pseudobombax grandiflorum* (Bombacaceae) by parakeets *Brotogeris versicolurus* (Psittacidae) in a forest fragment in Brazil. *Biotropica* 34:613–15.

Fuller, E. 2000. *Extinct Birds*. Oxford University Press, Oxford.
Gaban-Lima, R., M. A. Raposo, and E. Hofling. 2002. Description of a new species of *Pionopsitta* (Aves: Psittacidae) endemic to Brazil. *Auk* 119:815–19.
Galetti, M., and M. Rodrigues. 1992. Comparative seed predation on pods by parrots in Brazil. *Biotropica* 24:222–24.
Garnett, S., and G. Crowley. 1995. Feeding ecology of Hooded Parrots *Psephotus dissimilis* during the early wet season. *Emu* 95:54–61.
Garnett, S., G. Crowley, R. Duncan, N. Baker, and P. Doherty. 1993. Notes on live Night Parrot sightings in north-western Queensland. *Emu* 93:292–96.
Garnett, S. T., L. P. Pedler, and G. M. Crowley. 1999. The breeding biology of the Glossy Black-cockatoo *Calyptorhynchus lathami* on Kangaroo Island, South Australia. *Emu* 99:262–79.
Gastañaga, M., R. MacLeod, B. Hennessey, J. U. Núñez, E. Puse, A. Arrascue, J. Hoyos, W. M. Chambi, J. Vasquez, and G. Engblom. 2011. A study of the parrot trade in Peru and the potential importance of internal trade for threatened species. *Bird Conservation International* 21:76–85.
Gilardi, J. D., and C. A. Munn. 1998. Patterns of activity, flocking, and habitat use in parrots of the Peruvian Amazon. *Condor* 100:641–53.
Gonçalves da Silva, A., J. R. Eberhard, T. F. Wright, M. L. Avery, and M. A. Russello. 2010. Genetic evidence for high propagule pressure and long-distance dispersal in monk parakeet *(Myiopsitta monachus)* invasive populations. *Molecular Ecology* 19:3336–50.
González, J. A. 2003. Harvesting, local trade, and conservation of parrots in the Northeastern Peruvian Amazon. *Biological Conservation* 114:437–46.
Greene, T. C. 1998. Foraging ecology of the red-crowned parakeet *(Cyanoramphus novaezelandiae novaezelandiae)* and yellow-crowned parakeet *(C. auriceps auriceps)* on Little Barrier Island, Hauraki Gulf, New Zealand. *New Zealand Journal of Ecology* 22:161–71.
———. 1999. Aspects of the ecology of Antipodes Island Parakeet *(Cyanoramphus unicolor)* and Reischek's Parakeet *(C. novaezelandiae hochstetteri)* on Antipodes Island, October–November 1995. *Notornis* 46:301–10.
Harms, K. E., and J. R. Eberhard. 2003. Roosting behaviour of the Brown-throated Parakeet *(Aratinga pertinax)* and roost locations on four southern Caribbean islands. *Ornitologia Neotropical* 14:79–89.
Hausmann, F., K. E. Arnold, N. J. Marshall, and I. P. F. Owens. 2003. Ultraviolet signals in birds are special. *Proceedings of the Royal Society London B* 270:61–67.
Healy, C. 1990. *Maring Hunters and Traders: Production and Exchange in the Papua New Guinea Highlands*. University of California Press, Berkeley.
Heinrich, B., and T. Bugnyar. 2005. Testing problem solving in ravens: String-pulling to reach food. *Ethology* 111:962–76.
Heinsohn, R. 2008. The ecological basis of unusual sex roles in reverse-dichromatic eclectus parrots. *Animal Behaviour* 76:97–103.
Heinsohn, R., and M. Cermak. 2008. *Life in the Cape York Rainforest*. CSIRO Publishing, Melbourne.
Heinsohn, R., N. E. Langmore, A. Cockburn, and H. Kokko. 2011. Adaptive secondary sex ratio adjustments via sex-specific infanticide in a bird. *Current Biology* 21:1744–47.
Heinsohn, R., and S. Legge. 2003. Breeding biology of the reverse-dichromatic, co-operative parrot *Eclectus roratus*. *Journal of Zoology, London* 259:197–208.

Heinsohn, R., S. Legge, and J. A. Endler. 2005. Extreme reversed sexual dichromatism in a bird without sex role reversal. *Science* 309:617–19.

Heinsohn, R., S. Murphy, and S. Legge. 2003. Overlap and competition for nest holes among eclectus parrots, palm cockatoos and sulphur-crested cockatoos. *Australian Journal of Zoology* 51:81–94.

Heinsohn, R., T. Zeriga, S. Murphy, P. Igag, S. Legge, and A. Mack. 2009. Do Palm Cockatoos *(Probosciger aterrimus)* have long enough lifespans to support their low reproductive success? *Emu* 109:183–91.

Higgins, P. J., ed. 1999. *Handbook of Australian, New Zealand and Antarctic Birds.* Vol. 4: Parrots to Dollarbird. Oxford University Press, Melbourne.

Hile, A. G., N. T. Burley, C. B. Coopersmith, V. S. Foster, and G. F. Striedter. 2005. Effects of male vocal learning on female behaviour in the budgerigar, *Melopsittacus undulatus*. *Ethology* 111:901–23.

Hile, A. G., and G. F. Striedter. 2000. Call convergence within groups of female budgerigars *(Melopsittacus undulatus)*. *Ethology* 106:1105–14.

Hill, G. E. 2010. *National Geographic Bird Coloration*. National Geographic Society, Washington, DC.

Hingston, A. B., B. D. Gartrell, and G. Pinchbeck. 2004. How specialized is the plant-pollinator association between *Eucalyptus globulus* ssp. *globulus* and the swift parrot *Lathamus discolor? Austral Ecology* 29:624–30.

Homberger, D. G. 2006. Classification and status of wild populations of parrots. In *Manual of Parrot Behavior*, ed. A. U. Luescher, 3–11, Blackwell Publishing, Ames.

Hopper, S. D., and A. A. Burbidge. 1979. Feeding behaviour of a Purple-crowned Lorikeet on flowers of *Eucalyptus buprestium*. *Emu* 79:40–42.

Howe, H. F. 1980. Monkey dispersal and waste of a neotropical fruit. *Ecology* 61: 944–59.

Huber, L., and G. K. Gajdon. 2006. Technical intelligence in animals: The kea model. *Animal Cognition* 9:295–305.

Igag, P. 2002. The conservation of large rainforest parrots: A study of the breeding biology of Palm Cockatoos, Eclectus Parrots and Vulturine Parrots. MSc thesis, Australian National University.

Jones, C. G. 1987. The larger land-birds of Mauritius. In *Studies of Mascarene Island Birds*, ed. A. W. Diamond, 208–300, Cambridge University Press, Cambridge.

Jones, D. 2008. Feed the birds? *Wingspan* 18, no. 1: 16–19.

Joseph, L., G. Dolman, S. Donnellan, K. M. Saint, M. L. Berg, and A. T. D. Bennett. 2008. Where and when does a ring start and end? Testing the ring-species hypothesis in a species complex of Australian parrots. *Proceedings of the Royal Society B* 275:2431–40.

Joseph, L., A. Toon, E. E. Schirtzinger, and T. F. Wright. 2011. Molecular systematics of two enigmatic genera *Psittacella* and *Pezoporus* illuminate the ecological radiation of Australo-Papuan parrots (Aves: Psittaciformes). *Molecular Phylogenetics and Evolution* 59:675–84.

Juniper, T. 2002. *Spix's Macaw: The Race to Save the World's Rarest Bird*. Fourth Estate, London.

Juniper, T., and M. Parr. 1998. *Parrots: A Guide to the Parrots of the World*. Pica Press, Sussex.

Karubian, J., J. Fabara, D. Yunes, J. P. Jorgenson, D. Romo, and T. B. Smith. 2005. Temporal and spatial patterns of macaw abundance in the Ecuadorian Amazon. *Condor* 107:617–26.

Kistler. A. L., A. Gancz, S. Clubb, P. Skewes-Cox, K. Fischer, K. Sorber, C. Y. Chiu, A. Lublin, S. Mechani, Y. Farnoushi, A. Greninger, C. C. Wen, S. B. Karlene, D. Ganem, and J. L. DeRisi. 2008. Recovery of divergent avian bornaviruses from cases of proventricular dilation disease: Identification of a candidate etiologic agent. *Virology Journal* 5:88.

Krebs, E. A. 1998. Breeding biology of crimson rosellas *(Platycercus elegans)* on Black Mountain, Australian Capital Territory. *Australian Journal of Zoology* 46: 119–36.

———. 1999. Last but not least: Nestling growth and survival in asynchronously hatching crimson rosellas. *Journal of Animal Ecology* 68:266–81.

Krebs, E. A., and R. D. Magrath. 2000. Food allocation in crimson rosella broods: Parents differ in their responses to chick hunger. *Animal Behaviour* 59:739–51.

Laurance, W. F., M. Goosem and S. G. W. Laurance. 2009. Impacts of roads and linear clearings on tropical forests. *Trends in Ecology and Evolution* 24:659–69.

Legge, S., R. Heinsohn, and S. Garnett. 2004. Availability of nest hollows and breeding population size of eclectus parrots, *Eclectus roratus*, on Cape York Peninsula, Australia. *Wildlife Research* 31:149–61.

Lendon, A. H. 1973. *Australian Parrots in Field and Aviary* [originally written and illustrated by Neville W. Cayley]. Angus and Robertson, Sydney.

Lindsey, G. D., W. J. Arendt, J. Kalina, and G. W. Pendleton. 1991. Home range and movements of juvenile Puerto Rican parrots. *Journal of Wildlife Management* 55:318–22.

Low, R. 2005. *Amazon Parrots: Aviculture, Trade and Conservation.* DONA Publishing, Komemského.

Lowry, H., and A. Lill. 2007. Ecological factors facilitating city-dwelling in red-rumped parrots. *Wildlife Research* 34:624–31.

Mack, A. L., and D. D. Wright. 1996. Notes on occurrence and feeding of birds at Crater Mountain Biological Research Station, Papua New Guinea. *Emu* 96:89–101.

Magrath, R. D., and A. Lill. 1985. Age-related differences in behaviour and ecology of crimson rosellas, *Platycercus elegans*, during the non-breeding season. *Australian Wildlife Research* 12:299–306.

Marsden, S. J., M. Whiffin, L. Sadgrove, and P. Guimarães Jr. 2000. Parrot populations and habitat use in and around two lowland Atlantic forest reserves, Brazil. *Biological Conservation* 96:209–17.

Masello, J. F., T. Lubjuhn, and P. Quillfeldt. 2008. Is the structural and psittacofulvin-based coloration of wild burrowing parrots *Cyanoliseus patagonus* condition dependent? *Journal of Avian Biology* 39:653–62.

———. 2009. Hidden dichromatism in the Burrowing Parrot *(Cyanoliseus patagonus)* as revealed by spectrometric colour analysis. *Hornero* 24:47–55.

Masello, J. F., and P. Quillfeldt. 2002. Chick growth and breeding success of the Burrowing Parrot. *Condor* 104:574–86.

———. 2003. Body size, body condition and ornamental feathers of Burrowing Parrots: Variation between years and sexes, assortative mating and influences on breeding success. *Emu* 103:149–61.

———. 2004. Are haematological parameters related to body condition, ornamentation and breeding success in wild burrowing parrots *Cyanoliseus patagonus? Journal of Avian Biology* 35:445–54.

Matuzak, G. D., M. B. Bezy, and D. J. Brightsmith. 2008. Foraging ecology of parrots in a modified landscape: Seasonal trends and introduced species. *Wilson Journal of Ornithology* 120:353–65.

Mayr, E., and J. Diamond. 2001. *The Birds of Northern Melanesia: Speciation, Ecology, and Biogeography*. Oxford University Press, Oxford.

McGraw, K. J., and M. C. Nogare. 2005. Distribution of unique red feather pigments in parrots. *Biology Letters* 1:38–43.

Meehan, C., and J. Mench. 2006. Captive parrot welfare. In *Manual of Parrot Behavior*, ed. A. U. Luescher, 301–18, Blackwell Publishing, Ames.

Mettke-Hofmann, C., M. Wink, H. Winkler, and B. Leisler. 2004. Exploration of environmental changes relates to lifestyle. *Behavioral Ecology* 16:247–54.

Minnis, P. E., M. E. Whalen, J. H. Kelley, and J. D. Stewart. 1993. Prehistoric macaw breeding in the North American Southwest. *American Antiquity* 58:270–76.

Monterrubio, T., E. Enkerlin-Hoeflich, and R. B. Hamilton. 2002. Productivity and nesting success of Thick-billed Parrots. *Condor* 104:788–94.

Monterrubio-Rico, T. C., J. Cruz-Nieto, E. Enkerlin-Hoeflich, D. Venegas-Holguin, L. Tellez-Garcia, and C. Marin-Togo. 2006. Gregarious nesting behaviour of Thick-billed Parrots *(Rhynchopsitta pachyrhyncha)* in Aspen stands. *Wilson Journal of Ornithology* 118:237–43.

Moorhouse, R. J. 1997. The diet of the North Island kaka *(Nestor meridionalis septentrionalis)* on Kapiti Island. *New Zealand Journal of Ecology* 21:141–52.

Moorhouse, R., T. Greene, P. Dilks, R. Powlesland, L. Moran, G. Taylor, A. Jones, J. Knegtmans, D. Wills, M. Pryde, I. Fraser, A. August, and C. August. 2003. Control of introduced mammalian predators improves kaka *Nestor meridionalis* breeding success: Reversing the decline of a threatened New Zealand parrot. *Biological Conservation* 110:33–44.

Munshi-South, J., and G. S. Wilkinson. 2006. Diet influences lifespan in parrots (Psittaciformes). *Auk* 123:108–18.

Murphy, S. A. 2005. The ecology and conservation biology of Palm Cockatoos *Probosciger aterrimus*. PhD thesis, Australian National University.

Murphy, S., S. Legge, and R. Heinsohn. 2003. The breeding biology of palm cockatoos (*Probosciger aterrimus*): A case of a slow life history. *Journal of Zoology, London* 261:1–13.

Ndithia, H., M. R. Perrin, and M. Waltert. 2007. Breeding biology and nest site characteristics of the Rosy-faced Lovebird *Agapornis roseicollis* in Namibia. *Ostrich* 78:13–20.

Negro, J. J., J. H. Sarasola, F. Fariñas, and I. Zorrilla. 2006. Function and occurrence of facial flushing in birds. *Comparative Biochemistry and Physiology, Part A* 143:78–84.

Nichols, A. 2008. The complete fragments of Ctesias of Cnidus: Translation and commentary with an introduction. PhD thesis, University of Florida.

Olsen, P. 2007. *Glimpses of Paradise: The Quest for the Beautiful Parakeet*. National Library of Australia, Canberra.

———. 2009. Night parrots: Fugitives of the inland. In *Boom & Bust: Bird Stories for a Dry Country*, eds. L. Robin, R. Heinsohn, and L. Joseph, 121–46, CSIRO Publishing, Melbourne.

Olsen, P. D., J. Olsen, and I. Mason. 1993. Breeding and non-breeding season diet of the Peregrine Falcon *Falco peregrinus* near Canberra, prey selection, and the

relationship between diet and reproductive success. In *Australian Raptor Studies*, ed. P. D. Olsen, 55–77, Australasian Raptor Association, RAOU, Melbourne.
Pacheco, M. A., S. R. Beissinger, and C. Bosque. 2010. Why grow slowly in a dangerous place? Postnatal growth, thermoregulation, and energetics of nestling Green-rumped Parrotlets *(Forpus passerinus)*. *Auk* 127:558–70.
Pain, D. J., T. L. F. Martins, M. Boussekey, S. H. Diaz, C. T. Downs, J. M. M. Ekstrom, S. Garnett, J. D. Gilardi, D. McNiven, P. Primot, S. Rouys, M. Saoumoé, C. T. Symes, S. A. Tamungang, J. Theuerkauf, D. Villafuerte, L. Verfailles, P. Widmann, and I. D. Widmann. 2006. Impact of protection on nest take and nesting success of parrots in Africa, Asia and Australasia. *Animal Conservation* 9:322–30.
Patel, A. D., J. R. Iversen, M. R. Bregman, and I. Schulz. 2009. Experimental evidence for synchronization to a musical beat in a nonhuman animal. *Current Biology* 19:827–30.
Pearn, S. M., A. T. D. Bennett, and I. C. Cuthill. 2001. Ultraviolet vision, fluorescence and mate choice in a parrot, the budgerigar *Melopsittacus undulatus*. *Proceedings of the Royal Society London B* 268:2273–79.
———. 2003. The role of ultraviolet-A reflectance and ultraviolet-A induced fluorescence in the appearance of budgerigar plumage: Insights from spectrofluorometry and reflectance spectrophotometry. *Proceedings of the Royal Society London B* 270:859–65.
Pepper, J. W. 1996. The behavioural ecology of the Glossy Black-cockatoo *Calyptorhynchus lathami halmaturinus*. PhD thesis, University of Michigan.
Pepperberg, I. M. 1999. *The Alex Studies: Cognitive and Communicative Abilities of Grey Parrots*. Harvard University Press, Cambridge.
———. 2004. “Insightful” string-pulling in Grey parrots *(Psittacus erithacus)* is affected by vocal competence. *Animal Cognition* 7:263–66.
———. 2006. Grey Parrot cognition and communication. In *Manual of Parrot Behaviour*, ed. A. U. Luescher, 133–45, Blackwell Publishing, Ames.
———. 2007. Grey parrots do not always “parrot”: The roles of imitation and phonological awareness in the creation of new labels from existing vocalizations. *Language Sciences* 29:1–13.
———. 2008. *Alex & Me*. Scribe, Melbourne.
Plant, M. 2008. Good practice when feeding wild birds. *Wingspan* 18, no. 1: 20–23.
Powell, L. L., T. U. Powell, G. V. N. Powell, and D. J. Brightsmith. 2009. Parrots take it with a grain of salt: Available sodium content may drive collpa (clay lick) selection in southeastern Peru. *Biotropica* 41:279–82.
Ragusa-Netto, J. 2005. Extensive consumption of *Tabebuia aurea* (Manso) Benth. & Hook. (Bignoniaceae) nectar by parrots in a tecoma savanna in the southern Pantanal, Brazil. *Brazilian Journal of Biology* 65:339–44.
Renton, K. 2001. Lilac-crowned Parrot diet and food resource availability: Resource tracking by a parrot seed predator. *Condor* 103:62–69.
———. 2002. Seasonal variation in occurrence of macaws along a rainforest river. *Journal of Field Ornithology* 73:15–19.
———. 2004. Agonistic interactions of nesting and nonbreeding macaws. *Condor* 106:354–62.
Renton, K., and A. Salinas-Melgoza. 1999. Nesting behaviour of the Lilac-crowned Parrot. *Wilson Bulletin* 111:488–93.
———. 2004. Climatic variability, nest predation, and reproductive output of Lilac-

crowned Parrots *(Amazona finschi)* in tropical dry forests of western Mexico. *Auk* 121:1214–25.

Ribas, C. C., C. Y. Miyaki, and J. Cracraft. 2009. Phylogenetic relationships, diversification and biogeography in Neotropical *Brotogeris* parakeets. *Journal of Biogeography* 36:1712–29.

Ribas, C. C., R. G. Moyle, C. Y. Miyaki, and J. Cracraft. 2007. The assembly of montane biotas: Linking Andean tectonics and climatic oscillations to independent regimes of diversification in *Pionus* parrots. *Proceedings of the Royal Society B* 274:2399–2408.

Robbins, L. E. 2002. *Elephant Slaves and Pampered Parrots: Exotic Animals in Eighteenth-Century Paris*. Johns Hopkins University Press, Baltimore.

Rodríguez-Castillo, A. M., and J. R. Eberhard. 2006. Reproductive behavior of the Yellow-crowned Parrot (*Amazona ochrocephala*) in western Panama. *Wilson Journal of Ornithology* 118:225–36.

Rowley, I. 1980. Parent-offspring recognition in a cockatoo, the galah, *Cacatua roseicapilla*. *Australian Journal of Zoology* 28:445–56.

———. 1990. *Behavioural Ecology of the Galah* Eolophus roseicapillus *in the Wheatbelt of Western Australia*. Surrey Beatty and Sons, Sydney.

———. 1997. Family Cacatuidae (Cockatoos). In *Handbook of the Birds of the World*, eds. J. del Hoyo, A. Elliot, and J. Sargatal, vol. 4, Sandgrouse to Cuckoos, 246–79, Lynx Edicions, Barcelona.

Rowley, I., and G. Chapman. 1991. The breeding biology, food, social organisation, demography and conservation of the Major Mitchell or Pink Cockatoo, *Cacatua leadbeateri*, on the margin of the Western Australian wheatbelt. *Australian Journal of Zoology* 39:211–61.

Salinas-Melgoza, A., and K. Renton. 2007. Postfledging survival and development of juvenile lilac-crowned parrots. *Journal of Wildlife Management* 71:43–50.

Santos, S. I. C. O., B. Elward, and J. T. Lumeij. 2006. Sexual dichromatism in the blue-fronted Amazon parrot *(Amazona aestiva)* revealed by multiple-angle spectrometry. *Journal of Avian Medicine and Surgery* 20:8–14.

Saunders, D. A. 1982. The breeding behaviour and biology of the short-billed form of the White-tailed Black Cockatoo *Calyptorhynchus funereus*. *Ibis* 124:422–55.

———. 1983. Vocal repertoire and individual vocal recognition in the short-billed white-tailed black cockatoo, *Calyptorhynchus funereus latirostris* Carnaby. *Australian Wildlife Research* 10:527–36.

Saunders, D. A., G. T. Smith, and I. Rowley. 1982. The availability and dimensions of tree hollows that provide nest sites for cockatoos (Psittaciformes) in Western Australia. *Australian Wildlife Research* 9:541–56.

Saunders, D. L., and R. Heinsohn. 2008. Winter habitat use by the endangered, migratory Swift parrot (*Lathamus discolor*) in New South Wales. *Emu* 108:81–89.

Scarl, J. C., and J. W. Bradbury. 2009. Rapid vocal convergence in an Australian cockatoo, the galah *Eolophus roseicapillus*. *Animal Behaviour* 77:1019–26.

Schachner, A., T. F. Brady, I. M. Pepperberg, and M. D. Hauser. 2009. Spontaneous motor entrainment to music in multiple vocal mimicking species. *Current Biology* 19:831–36.

Schuck-Paim, C., W. J. Alonso, and E. O. Ottoni. 2008. Cognition in an ever-changing world: Climatic variability is associated with brain size in Neotropical parrots. *Brain, Behavior and Evolution* 71:200–215.

Schweizser, M., O. Seehausen, M. Güntert, and S. T. Hertwig. 2010. The evolutionary diversification of parrots supports a taxon pulse model with multiple trans-oceanic dispersal events and local radiations. *Molecular Phylogenetics and Evolution* 54:984–94.

Seixas, G. H. F., and G. de M. Mourão. 2002. Nesting success and hatching survival of the Blue-fronted Amazon (*Amazona aestiva*) in the Pantanal of Mato Grosso do Sul, Brazil. *Journal of Field Ornithology* 73:399–409.

Shwartz, A., D. Strubbe, C. J. Butler, E. Matthysen, and S. Kark. 2009. The effect of enemy-release and climate conditions on invasive birds: A regional test using the rose-ringed parakeet *(Psittacula krameri)* as a case study. *Diversity and Distributions* 15:310–18.

Smith, G. T. 1991. Breeding ecology of the western long-billed corella, *Cacatua pastinator pastinator*. *Wildlife Research* 18:91–110.

Smith, J., and A. Lill. 2008. Importance of eucalypts in exploitation of urban parks by Rainbow and Musk Lorikeets. *Emu* 108:187–95.

Snyder, N. F. R. 2004. *The Carolina Parakeet: Glimpses of a Vanished Bird*. Princeton University Press, Princeton.

Snyder, N. F. R., J. W. Wiley, and C. B. Kepler. 1987. *The Parrots of Luquillo: Natural History and Conservation of the Puerto Rican Parrot*. Western Foundation of Vertebrate Zoology, Los Angeles.

Snyder, N., P. McGowan, J. Gilardi, and A. Grajal., eds. 2000. *Parrots: Status Survey and Conservation Action Plan 2000–2004*. IUCN, Gland.

Somerville, A. D., B. A. Nelson, and K. J. Knudson. 2010. Isotopic investigation of pre-Hispanic macaw breeding in Northwest Mexico. *Journal of Anthropological Archaeology* 29:125–35.

South, J. M., and S. Pruett-Jones. 2000. Patterns of flock size, diet, and vigilance of naturalized Monk Parakeets in Hyde Park, Chicago. *Condor* 102:848–54.

Steadman, D. W. 2006. *Extinction and Biogeography of Tropical Pacific Birds*. University of Chicago Press, Chicago.

Stoleson, S. H., and S. R. Beissinger. 2001. Does risk of nest failure or adult predation influence hatching patterns of the Green-rumped Parrotlet? *Condor* 103:85–97.

Strubbe, D., and E. Matthysen. 2009. Establishment success of invasive ring-necked and monk parakeets in Europe. *Journal of Biogeography* 36:2264–78.

———. 2009. Experimental evidence for nest-site competition between invasive ring-necked parakeets *(Psittacula krameri)* and native nuthatches *(Sitta europaea)*. *Biological Conservation* 142:1588–94.

Symes, C. T., and M. R. Perrin. 2003. Seasonal occurrence and local movements of the grey-headed (brown-necked) parrot *Poicephalus fusciollis suahelicus* in southern Africa. *African Journal of Ecology* 41:299–305.

Taysom, A. J., D. Stuart-Fox, and G. C. Cardoso. 2011. The contribution of structural-, psittacofulvin- and melanin-based colouration to sexual dichromatism in Australasian parrots. *Journal of Evolutionary Biology* 24:303–13.

Tidemann, S., and T. Whiteside. 2010. Aboriginal stories: The riches and colour of Australian birds. In *Ethno-ornithology: Birds, Indigenous Peoples, Culture and Society*, eds. S. Tidemann and A. Gosler, 153–79, Earthscan, London.

Tracey, J., M. Bomford, Q. Hart, G. Saunders, and R. Sinclair. 2007. *Managing Bird Damage to Fruit and Other Horticultural Crops*. Bureau of Rural Sciences, Canberra.

Trivedi, M. R., F. H. Cornejo, and A. R. Watkinson. 2004. Seed predation on Brazil

Nuts *(Bertholletia excelsa)* by Macaws (Psittacidae) in Madre de Dios, Peru. *Biotropica* 36:118–22.

Trounson, D., and M. Trounson. 1998. *Australian Birds Simply Classified.* 4th ed. Murray David Publishing, Sydney.

Vicentini, A., and E. A. Fischer. 1999. Pollination of *Moronobea coccinea* (Clusiaceae) by the Golden-winged Parakeet in the central Amazon. *Biotropica* 31:692–96.

Waltman, J. R., and S. R. Beissinger. 1992. Breeding behavior of the Green-rumped Parrotlet. *Wilson Bulletin* 104:65–84.

Wanker, R., J. Apcin, B. Jennerjahn, and B. Waibel. 1998. Discrimination of different social companions in spectacled parrotlets *(Forpus conspicillatus):* Evidence for individual vocal recognition. *Behavioral Ecology and Sociobiology* 42:197–202.

Wanker, R., L. Cruz Bernate, and D. Franck. 1996. Socialization of Spectacled Parrotlets *Forpus conspicillatus*: The role of parents, crèches and sibling groups in nature. *Journal für Ornithologie* 137:447–61.

Wanker, R., Y. Sugama, and S. Prinage. 2005. Vocal labelling of family members in spectacled parrotlets, *Forpus conspicillatus. Animal Behaviour* 70:111–18.

Webb, H. P. 1997. Nesting and other observations of Solomon Island birds. *Australian Bird Watcher* 17:34–41.

Webb, N. V., R. F. Thomas, and J. R. Millam. 2010. The effect of rope color, size and fray on environmental enrichment device interaction in male and female Orange-winged Amazon parrots *(Amazona amazonica). Applied Animal Behaviour Science* 124:149–56.

Westcott, D. A., and A. Cockburn. 1988. Flock size and vigilance in parrots. *Australian Journal of Zoology* 36:335–49.

White, N. E., M. J. Phillips, M. T. P. Gilbert, A. Alfaro-Núñez, E. Willerslev, P. R. Mawson, P. B. S. Spencer, and M. Bunce. 2011. The evolutionary history of cockatoos (Aves: Psittaciformes: Cacatuidae). *Molecular Phylogenetics and Evolution* 59:615–62.

Whitehead, J. K. 2007. Breeding success of adult kakapo *(Strigops habroptilus)* on Codfish Island (Whenua Hou): Correlations with foraging home ranges and habitat selection. PhD thesis, Lincoln University.

Williams, M. ed. 2006. A Celebration of Kakapo [Special Issue]. *Notornis* 53, no. 1.

Wirminghaus, J. O., C. T. Downs, M. R. Perrin, and C. T. Symes. 2001. Abundance and activity patterns of the Cape parrot *(Poicephalus robustus)* in two afromontane forests in South Africa. *African Zoology* 36:71–77.

Wright, T. F. 1996. Regional dialects in the contact call of a parrot. *Proceedings of the Royal Society of London B* 263:867–72.

Wright, T. F., and C. R. Dahlin. 2007. Pair duets in the Yellow-naped amazon *(Amazona auropalliata):* Phonology and syntax. *Behaviour* 144:207–28.

Wright, T. F., C. R. Dahlin, and A. Salinas-Melgoza. 2008. Stability and change in vocal dialects of the Yellow-naped amazon. *Animal Behaviour* 76:1017–27.

Wright, T. F., E. E. Schirtzinger, T. Matsumoto, J. R. Eberhard, G. R. Graves, J. J. Sanchez, S. Capelli, H. Müller, J. Scharpegge, G. K. Chambers, and R. C. Fleischer. 2008. A multilocus molecular phylogeny of the parrots (Psittaciformes): Support for a Gondwanan origin during the Cretaceous. *Molecular Biology and Evolution* 25:2141–56.

Wright, T. F., C. A. Toft, E. Enkerlin-Hoeflich, J. Gonzalez-Elizondo, M. Albornoz, A. Rodríguez-Ferraro, F. Rojas-Suárez, V. Sanz, A. Trujillo, S. R. Beissinger,

V. Berovides, A. X. Gálvez A., T. Brice, K. Joyner, J. Eberhard, J. Gilardi, S. E. Koenig, S. Stoleson, P. Martuscelli, J. M. Meyers, K. Renton, A. M. Rodríguez, A. C. Sosa-Asanza, F. J. Vilella, and J. W. Wiley. 2001. Nest poaching in Neotropical parrots. *Conservation Biology* 15:710–20.

Wyndham, E. 1981. Breeding and mortality of Budgerigars *Melopsittacus undulatus*. *Emu* 81:240–43.

———. 1982. Movements and breeding seasons of the Budgerigar. *Emu* 82:276–82.

Index

Page numbers in **bold** refer to illustrations. Page numbers in *italics* refer to tables and figures.

DATE DUE